Ecological Investigations

These investigations identify and clarify some basic assumptions and methodological principles involved in ecological explanations of plant associations. How are plants geographically distributed into characteristic groups? What are the basic conditions that organize groups of interspecific plant populations that are characteristic of particular kinds of habitats? Answers to these questions concerning the geographical distribution of plants in late 19th century European plant geography and early 20th century American plant ecology can be distinguished according to differing logical assumptions concerning the habitats of plant associations. Through an analysis of several significant case studies in the early history of plant ecology, Konopka distinguishes a logic of habitats that conceives of plant associations through an analogy with individual organisms from a logic in which plant associations have a reciprocal relation to habitat physiography. He argues that a phenomenological conception of the logical attributes of habitats can philosophically complement the physiographic tradition in early plant ecology and provide an attractive alternative to standard reductionism and holism debates that persist today. This wide ranging and original analysis will be valuable for readers interested in the history and philosophy of ecology.

Adam C. Konopka teaches in the philosophy department at Xavier University in Cincinnati, Ohio (USA). His research interests are primarily in phenomenology and the philosophy of science. He has published several articles in journals such as the *New Yearbook in Phenomenology and Phenomenological Philosophy*, *Ethics, Policy, Environment*, and *Environmental Ethics*.

History and Philosophy of Biology

Series Editor: Rasmus Grønfeldt Winther is Associate Professor of Philosophy at the University of California, Santa Cruz (UCSC).

This series explores significant developments in the life sciences from historical and philosophical perspectives. Historical episodes include Aristotelian biology, Greek and Islamic biology and medicine, Renaissance biology, natural history, Darwinian evolution, nineteenth-century physiology and cell theory, twentieth-century genetics, ecology, and systematics, and the biological theories and practices of non-Western perspectives. Philosophical topics include individuality, reductionism and holism, fitness, levels of selection, mechanism and teleology, and the nature-nurture debates, as well as explanation, confirmation, inference, experiment, scientific practice, and models and theories vis-à-vis the biological sciences.

Authors are also invited to inquire into the "and" of this series. How has, does, and will the history of biology impact philosophical understandings of life? How can philosophy help us analyze the historical contingency of, and structural constraints on, scientific knowledge about biological processes and systems? In probing the interweaving of history and philosophy of biology, scholarly investigation could usefully turn to values, power, and potential future uses and abuses of biological knowledge.

The scientific scope of the series includes evolutionary theory, environmental sciences, genomics, molecular biology, systems biology, biotechnology, biomedicine, race and ethnicity, and sex and gender. These areas of the biological sciences are not silos, and tracking their impact on other sciences such as psychology, economics, and sociology, and the behavioral and human sciences more generally, is also within the purview of this series.

Philosophy of Biology Before Biology
Edited by Cécilia Bognon-Küss & Charles T. Wolfe

Ecological Investigations
A Phenomenology of Habitats
Adam C. Konopka

For more information about this series, please visit: https://www.routledge.com/History-and-Philosophy-of-Biology/book-series/HAPB

Ecological Investigations
A Phenomenology of Habitats

Adam C. Konopka

Routledge
Taylor & Francis Group

LONDON AND NEW YORK

First published 2020
by Routledge
2 Park Square, Milton Park, Abingdon, Oxon OX14 4RN

and by Routledge
605 Third Avenue, New York, NY 10017

First issued in paperback 2021

Routledge is an imprint of the Taylor & Francis Group, an informa business

British Library Cataloguing-in-Publication Data
A catalogue record for this book is available from the British Library

Library of Congress Cataloging-in-Publication Data
Names: Konopka, Adam, author.
Title: Ecological investigations: a phenomenology of habitats /
Adam C. Konopka.
Description: Milton Park, Abingdon, Oxon; New York,
NY: Routledge, 2020. |
Series: History and philosophy of biology | Includes bibliographical
references and index.
Identifiers: LCCN 2019017795 | ISBN 9781138300378 (hbk) |
ISBN 9780203733523 (ebk)
Subjects: LCSH: Plant ecology–America–History–20th century. |
Plant ecology–Europe–History–19th century. |
Biogeography–America–History–20th century. |
Biogeography–Europe–History–19th century.
Classification: LCC QK109 .K66 2020 | DDC 581.7–dc23
LC record available at https://lccn.loc.gov/2019017795

ISBN 13: 978-0-367-78432-4 (pbk)
ISBN 13: 978-1-138-30037-8 (hbk)

Typeset in Times New Roman
by Deanta Global Publishing Services, Chennai, India

For Louis-Michael, Devon Haskell, and Judy Ramona

Contents

Acknowledgments

I would like to acknowledge the support that made this work possible. This book was written while I held a visiting position at Xavier University in Cincinnati, Ohio, that afforded a generous research schedule and an opportunity to teach small undergraduate seminars. I am thankful to the students of these seminars for their collaboration in the learning process and for the hospitality of Xavier's philosophy department. I have also worked through these investigations at meetings of the Husserl Circle, Society of Phenomenological and Existential Philosophy, University of Kentucky Philosophy Department, Phenomenology Research Group at Loyola University Chicago, and International Association for Environmental Philosophy. I received valuable feedback and helpful direction from colleagues at the meetings of these organizations. This work was also stimulated by conversations with several colleagues, especially Charles S. Brown, Tim Brownlee, Brett Buchanan, Sam Cocks, John Drummond, Gabe Gottlieb, Hanne Jacobs, Bob Sandmeyer, Kathleen Smythe, Mike Sontag, David Story, and Ron White. Richard Polt and Dan Dwyer read a draft of the manuscript and provided helpful comments and feedback. This work was also supported through personal relationships with my family and friends. I would like to thank my wife, Erica, for her inspiration and loving support and my children – Louis-Michael, Devon Haskell, and Judy Ramona – for sharing their sense of wonder for the natural world with me.

An earlier version of Chapter Four was published as "Environed Embodiment and Geometric Space" in *Phenomenology and Place*, edited by Janet Donohoe (Lanham: Rowman and Littlefield, 2017), 143–161.

Introduction

On the empirical and logical foundations of ecology

These investigations are about how habitats shape the way that some plants typically grow together. Many gardeners can attest to a common observation that certain plants grow well together in a way that is conditioned by a variety of factors, e.g., plant growth structure (stalk, stem, and vine), light exposure, soil and mineral content, and temperature tolerance. Many hikers can attest to a common observation that the composition of plants changes according to changes in altitude and geography. Many travelers can attest to a common observation that similar groupings of plants grow in places that are far away from each other. These common observations can be a source of wonder that motivates an interest in asking the question – how are plants geographically distributed? How do habitats condition the way plants grow together?

The question concerning the geographical distribution of plants can arise in ordinary contexts of gardening, hiking, and traveling. This question is also a unique point of entry into some of the basic assumptions and explanatory principles in the science of plant ecology. These investigations explore how this question emerges for the first time as a scientific question in the 19th century and became the central question of the early plant ecologists. Through an exploration of the philosophical differences that underlie the various answers to this question, I develop my own approach and propose some philosophical tools that I think are helpful and even necessary for a propositional understanding of plant associations.

This question is an empirical question with scientific explanations that have been refined for over a century. While these explanations have given rise to further inquiry into other subject matter, the central question of the early plant ecologists highlights the unique relationship between empirical field research and logical classifications involved in theoretical modeling. All the sciences employ logical reasoning in some important sense and plant ecology is a particular and peculiar case in point. Plant ecology is a peculiar science in which observations of empirical plant collectives and the logical classifications of them are tightly joined. While the question concerning the geographical distribution of plants demands empirical answers, these answers inevitably and necessarily have basic assumptions and explanatory principles that can be logically clarified in a way that highlights the differences among them. The role of logic here is not to take the lead, so to speak, but to provide a second-order and supportive role that can assist the field researcher in the search for empirical answers to ecological questions. I think that the science of plant ecology not only needs robust experimental and field research agendas, on the one hand, but careful and critical conceptual analysis of theoretical principles, on the other hand. This dual explanatory necessity does not imply a hidden dualistic assumption regarding plant associations, but a radical empiricism with a second order justification that can contribute to a propositional understanding of ecological things.

Philosophical assumptions and explanatory principles concerning plant associations can be clearly illustrated in 19th century European plant geography and early 20th century American plant ecology. This early history illustrates at least two different sets of assumptions and principles concerning the logic of habitat associations. On the one hand, the forms proper to plant associations were conceived in an analogy with the organic forms of individual plants. This conception of the form proper to plant associations can be illustrated through Alexander von Humboldt's notion of "plant form" and Frederick Clements account of prairie succession. The suppositions concerning these notions of form proper to plant associations find their justification in an epistemological idealism that roots the necessary unities proper to plant associations in the cognitive achievements of the knower. On the other hand, the forms proper to plant associations were conceived in a physiographic logic of habitat associations that conceived of plant associations in a reciprocal or mutual relation with the geological and topographic features of a habitat. This conception of form proper to plant associations can be illustrated through Eugene Warming's notion of "growth form" and Henry Chandler Cowles' account of dune succession. The suppositions concerning these notions of form proper to plant associations find their justification in an epistemological realism that roots the necessary unities proper to plant associations in the role of water variations in terrestrial plant abundance and distribution. These two sets of assumptions and principles clarify the logical differences between epistemological idealisms and realisms in the history of plant ecology.

These investigations can also be generally characterized as a critique of uncritical epistemological idealisms in the philosophy of ecology. In contrast to the epistemological idealisms that ultimately account for the unity proper to ecological things in terms of the synthetic achievements of the cognizing knower, I propose a phenomenological account of ecological things that operates with a logic of habitat associations that is particular to geographical places. Generally speaking, I argue that a phenomenological philosophy can bolster the physiographic tradition in plant ecology by clarifying the logic of habitat fitness in its conception of plant associations and provide some methodological resources that are indispensable for the philosophy of ecology today. It is often said that ecological wholes are more than the sum of their parts; the following investigations attempt to clarify the senses of "more than the sum" in this claim. Philosophical investigation into ecological parts and wholes involves a reformulation of this slogan that prioritizes logical categories, e.g., unified definite manifolds and a logic of part-whole fitness, over the numeric conception of "sum" and conceives of "more" as immanent to particular habitats. This holistic claim can thus be reformulated –ecological wholes are not reducible to unorganized collections of ecological parts. I argue that a physiographic justification of this claim provides a helpful corrective to the tendency to idealize ecological wholes and, moreover, that a phenomenological account of unified definite manifolds and part-whole logics of fitness are formal tools that can effectively address some of the philosophical issues that are peculiar to ecological sciences.

Chapter One investigates three cases studies of succession in early 20th century American plant ecology. The accounts of succession involved in these classic case studies operated, in particular, with different logics of habitat associations. Cowles' account of dune succession, Clements' account of prairie succession, and Lindeman's account of lake succession – each of these contributions to the theory of succession in 20th century plant ecology have identifiable logics of part-whole relations that are presupposed by their classification of plant associations. What part-whole relations are involved in the assumptions regarding plant associations in these various accounts of succession? First, the part-whole logic that is supposed by Cowles' account of succession involved a mutual or reciprocal founding relationship between the topological features of sand dunes and the physiological features of pioneering plant populations, e.g., high rate of vertical shoot growth, rhizomic root structure, and low moisture tolerance. The identification of Cowles' logic of dune habitat associations clarifies the directional attributes at work in his account of succession as oriented toward a physiographic base level, e.g., topological variations involved in erosion and sediment deposition. Second, Clements' account of prairie succession operated with a logic of habitat associations that attributed functional and directional attributes to the association as a whole. That is, Clements' logic of habitat associations included and emphasized the functional interactions of plant populations as parts of community with biological attributes proper to itself in an analogous way in which organs are functional parts in the activity of individual organisms. Clements' account of prairie succession operated with a conception of plant association that had emergent organic properties that contribute a directionality to the sequence of prairie succession. Raymond Lindeman's logic of habitat associations examined the abundance of plant populations in terms of annual rates in productivity and, in particular, the chemical constituents of nutritive cycles. Lindeman's trophic dynamic account of lake succession transformed questions concerning the successive distribution of plants through a thematization of the variations among seasonal nutrient cycling. The study of lake succession in early 20th century limnology involved a shift in thematic focus – the new object of inquiry became the annual nutrient cycle. Lindeman's account of the late stages of lake succession on Cedar Bog Lake shifts the thematic focus from the distribution of plant populations in terms of phenotypic attributes to a thematization of the distribution of calories through an annual trophic cycle. His account of lake succession also illustrates a logic of habitat associations that takes nutrient productivity as the basic variation of succession through an abstraction to the chemical constituents of caloric transfer rates among trophic levels.

Chapter Two identifies and clarifies the logical difference between Humboldt's physiognomic and Warming's ecological plant geographies. Humboldt's pioneering logic of habitat associations supposed a botanical classification of regional plant associations, a Post-Kantian conception of organic form, and a geometric conception of spatial distribution. More specifically, Humboldt's physiognomic logic of habitat fitness prioritized the synthetic achievements of the imagination in accounting for the unity proper to individual plants and plant associations. I argue that Humboldt's part-whole logic and implicit theory of manifolds can be properly characterized as an epistemological idealism that relies on a one-sided emphasis on the synthetic achievements of the plant geographer to account for the unity proper to plant formations. In contrast, Warming's logic of habitat associations illustrates a nutritive conception of plant forms and a different theory of manifolds. Warming's logic of habitat fitness relied on a nutritive conception of plant forms that was an application of discoveries in 19th century plant physiology concerning the role of water in the process of photosynthesis. Warming's pioneering application of a nutritive conception of plant forms to questions concerning the geographical distribution of plant associations can be illustrated through his notion of "growth form" which he develops in direct contrast with the Humboldtian tradition in 19th century plant geography. More specifically, Warming operated with a theory of manifolds that thematized water variations as the basic identity among the manifolds of geographic and nutritive changes. This conception of unity in manifolds of plant geographical change can be properly characterized as an epistemological realism that prioritized the ideographic particularity of given habitats. This investigation reconstructs Warming's logic of habitat fitness and illustrates how this breakthrough was a departure from the fundamental assumptions and explanatory principles of 19th century plant geography.

The problem of habitat fitness is evident in the debates concerning succession in 20th century plant ecology. The part-whole relations that characterize a collection of organisms and their relation to their environment as internally unified, not externally related, also raises philosophical questions concerning the kinds of unity that are attributed to plant collectives. I argue that phenomenological resources are indispensable in answering these philosophical questions. In the later investigations, I develop applications of some of these resources through a contrast with Immanuel Kant's account of organic form (Chapter Three) and the conception of necessity in his conception of the *a priori* synthetic judgments (Chapter Four), a Newtonian account of absolute space (Chapter Five).

The dialogue with Kant in these following investigations has a variety of historical motivations that can help untangle some of the elements involved in the problem of habitat fitness. The first historical motivation concerns Kant's influence in the development of 19th century life sciences as an important background for the subsequent development of plant ecology, e.g., Alexander von Humboldt's plant geography. My genealogy of 19th century plant geography illustrates an epistemological tension between Humboldt's notion of "plant form" and Warming's notion of "growth form" (Chapter Two). I argue that this tension can be clarified through the different logics of habitat associations in Humboldt and Warming, a logical difference that persisted in 20th century American debates concerning plant succession between the Chicago and Nebraska schools' respective accounts of plant succession. Generally speaking, Kant's critical project provided epistemological resources for the organicist tradition in late 18th and early 19th century German biology. Humboldt's account of the forms proper to plant associations belongs in this tradition. In particular, Kant's account of organic form and his distinction between determinate and reflective judgment provided a philosophical framework for the subsequent organicist tradition in German biology and Humboldt, in particular. In short, my dialogue with Kant in these investigations is also a dialogue with some of the fundamental assumptions and explanatory principles in Humboldt's plant geography. The remaining investigations can even be framed in terms of a problematization of Humboldt's complex notion of plant associations that is comprised of a botanical notion of form as sensible shape, an organic notion of form in the Kantian sense, and geometric notion spatial distribution in the Newtonian sense. Through an immanent critique of this complex notion of plant associations in the remaining investigations, I identify and clarify several basic differences from my own phenomenological approach.

Chapter Three investigates the relationship between Kant's critical project and Edmund Husserl's phenomenology concerning the unity and functional attributes proper to biological individuals and collectives. While Husserl did not substantively engage the life sciences of his day, there is logical space in his philosophy of science that provides a distinctive alternative to the Kantian approach concerning how the unity of plant individuals and collectives are known to be organized. While a systematic study of Husserl's phenomenological alternative to Kant is beyond the scope of these investigations, an engagement with Kant on the following philosophical issues can highlight distinct differences between the two epistemological approaches:

1) Biological Form – unified definite manifold organized by parts and wholes that are characterized by a logic of habitat fitness (Chapter Three),
2) Necessity – the unity of manifold variations (Chapter Four), and
3) Place – founding relations between habituated environment and geometric space (Chapter Five).

Phenomenological accounts of these issues are different from Kantian and Post-Kantian accounts. The following investigations explore those differences through a dialogue with Kant and, in the process, develop logical resources that can clarify an ecological notion of form that is characterized by a logic of habitat fitness. Husserl and Kant provide different accounts of how the unity of objects in general is known and this difference is especially acute concerning organisms. For Kant, the unity of organic wholes is conditioned by and finds its justification in the unity of apperception – the "I think" that accompanies all my representations. For Husserl, by contrast, the unity of objects in general is conditioned by and finds its justification in the sense of the object itself as an identity-in-a-manifold. Husserl is an epistemological realist here in a way that Kant is not; the reasoning is not from the condition to the conditioned (as it is in Kant), but from the conditioned to the condition. I argue that this object-oriented justification has broad ramifications for how we come to know individual plants and plant collectives.

Chapter Four investigates how Husserl and Kant differ with regard to the necessity involved in *a priori* synthetic judgments. For Kant, *a priori* necessity is the opposite of contingency and is one of the three categories of modality (along with existence and possibility). Modal judgments have a "peculiar function" in that they determine the "value of the copula" in a proposition with respect to "thought in general."[1] The category of necessity is particularly significant for Kant's epistemology in that it (along with universality) is a determining feature of *a priori* knowledge that is independent from particular experiences. By contrast, Husserl radicalized and pluralized Kant's conception of the material *a priori* – distinguishing among the correlational *a priori*, *a priori* bound to the empirical, and pure material *a priori*. In short, this more nuanced account of necessity that Husserl initially develops in his logic of parts and wholes and finds its full justification in his mature theory of intentionality, provides an attractive alternative to the idealism in Kant's account of necessity. It also provides a logical alternative to the persistent idealist tendency in plant ecology toward uncritical holism that characterizes all ecological communities as organisms of a higher order through an analogy with individual organisms.

Chapter Five investigates the differences between phenomenological and Newtonian conceptions of space. I argue that the orientation awareness involved in embodied perception can be distinguished from geometric space and that there is a one-sided founding relationship between an environment and a Newtonian conception of absolute space. More specifically, my claim is that the internal relation between pre-reflective bodily self-awareness and environing world (*Umwelt*) has a founding relation with the concept of geometric space and that this relation is not reciprocal or mutual. This investigation identifies and preliminarily clarifies several continua (unified definite manifolds) involved in what I call "environed embodiment" that contribute to awareness of bodily location and movement. These continua play an important role in the justification of the constitutive priority of a local (in)determinate place in relation to geometric conceptions of space that 1) are emptied of material plenum, 2) have positional uniformity, and 3) involve an abstraction proper to the idealization involved in geometric figure and algebraic symbolization.

There are several potential objections that could be raised against the general theoretical thrust of these investigations. Briefly highlighting these objections and providing my preliminary responses to them affords an opportunity to situate and clarify the general approach of these investigations. These objections and responses also help motivate a concern for some of the philosophical issues involved in the history and philosophy of plant ecology. The first objection concerns an investigative over-emphasis on short-term habitat constraints that might have the appearance of a Lamarckian conception of adaptation. In other words, my investigations into the logic of habitat fitness of the early plant ecologists and problematization of Humboldt's conception of plant form suffers from a

Lamarckian view of biological adaptation that emphasizes the role of the abiotic habitat in the distribution of populations of plant species. According to the line criticism in this objection, my phenomenological investigations into the logic of habitat associations and conceptions of plant forms of the early plant ecologists operates with a thinly disguised version of a Lamarckian theory of heritable traits. The individualism–holism debates of the early 20th century American plant ecologists suffered from a Lamarckian conception of evolution that has been overcome through the methodological synthesis of evolutionary biology and ecosystems ecology in the later 20th century by authors such as Eugene Odum. These investigations are uninteresting not only in their backwards orientation that ignores the scientific progress in plant ecology, but celebrates past problems that have already been resolved by contemporary approaches. These investigations beat one of the dead horses in the history of plant ecology one more time and neglect contemporary topics and research agendas.

My response to this objection involves some basic methodological clarifications and a defense against the charge that my investigations are Lamarckian. Allow me to first briefly characterize Lamarck's conception of the evolution of heritable traits before distinguishing my phenomenological approach from it. Lamarckian inheritance is the hypothesis – often associated with Jean-Baptiste Lamarck (1744–1829) – that an organism passes on phenotypic traits that have changed due to environmental pressures to its offspring. Otherwise known as the "inheritance of acquired characteristics" or "soft inheritance," this hypothesis maintained that variations in phenotypic traits were acquired in adaptive responses to environmental variations and transmitted to successive generations. The reasoning here is from the environing habitat, to the behavior and phenotypic trait of the organism, to short-term intergenerational heritage. Darwin's theory of evolution by natural selection, by contrast, involved a difference sequence of explanations. Rather than the plant adapting to habitat variations, Darwinian adaptation occurs as a result of biological processes, e.g., interspecific competition, that contribute to reproductive fitness. The reasoning here is from the phenotypic traits of the organism, to long-term inheritance, and the carrying capacity variations. In short, the charge of a latent Lamarckianism to my investigations stems from my over-emphasis on habitat variations at the expense of the dynamic biological processes such as interspecific competition that govern speciation.

I have a few initial responses to this objection that can also help further situate the following investigations. The first is to distinguish the kinds of questions that are at work in evolutionary biology and plant ecology. The evolutionary biologist is interested in the diversity of species and asks the basic question: why are there so many kinds of organisms? Theories of evolution – whether Lamarckian, Darwinian, genetic, and so on – attempt to explain the causal origins of this diversity through evolutionary processes. The various theories of heritability can be distinguished in various ways, e.g., temporal scale (short- or long-term inheritance) and emphasis on biological or habitat variations, but they all stem from a basic interest in the causal origins of the heterogeneity of organisms. The plant ecologists of the late 19th and 20th centuries were asking a different basic question.

They were asking Humboldt's question concerning the geographical distribution of plants and addressing it through accounts of succession. Why do we find characteristic plant associations, e.g., beech-maple forests, prairie grasses, and white cedar-tamarack bogs, in similar geographical habitats? Whereas an interest in the diversity of organisms motivates a search for ultimate explanations through an appeal to historical origins, an interest in plant distributions motivates a search for proximate explanations through an appeal to habitat constraints. These are two different questions with different kinds of explanations. A plant ecologist, for example, could ask the question concerning the geographical distribution of plants without committing to a particular theory of heredity. In short, my investigations into the history and philosophy of plant ecology are not Lamarckian because they are largely agnostic toward theories of heritability. Rather, they are concerned with the distribution of plant associations that characteristically grow together, rather than the differences between plant species. The focus here is on spatial distribution, not phenotypic or genetic differences among species.

The second aspect of my response to the charge of a latent Lamarckian theory of inheritance concerns a clarification of timescales. Lamarckian adaptation is concerned with the direct interaction between an organism and its environment and the short-term heritability of traits to proximate generations of offspring. By contrast, the timescales of the logics of habitat associations of the late 19th and early 20th century plant ecologists were long-term processes involved in the surface geology, topology, and climate change (especially post-glacial). The timescales of the early plant ecologists were primarily geological, not biological. Proximate explanations of the direct interaction of plants with their environment are relevant as static profiles in dynamic long-term processes at work in geological and climatic history of the planet. In short, not only can the distributive questions of the early plant geographers and ecologists be distinguished from questions concerning the causal origins of the diversity of organisms, e.g., theories of inheritance, but the timescales of the logics of habitat associations that I explore in these investigations are of a different order than the adaptations that are passed on to the immediate generations in Lamarck's account.

Another potential objection to these investigations arises from historians who are suspicious of scholarship in the history of science generally, and the history of ecological sciences, in particular. According to this objection, these genealogical investigations are misguided in their theory of historiography and conception of the purpose of historical scholarship. More specifically, these historiographies suffer from a whiggish conception of the history of science that "studies the past with reference to the present."[2] A whiggish historiography of science is one that interprets scientific history as progressively improving toward its present self-realization. Various theories and scientific accomplishments are evaluated by the historiographer in light of the present scientific paradigm in which the historian writes and picks winners and losers according to whether they adhere and bolster the current scientific orthodoxy. Not only do whiggish historiographies suffer from a latent teleological view of history, but they do not sufficiently attend to the historical specificity and complexity of the scientific past. This charge of whiggish historiography can be pointed more directly at my investigations. These genealogies presume to reconstruct the part-whole logic of the classification systems of late 19th and early 20th century ecologists and identify the operative basic assumptions and explanatory principles, all the while ignoring the psychological, sociological, and political contexts in which they occurred. In short, these whiggish historiographies of plant ecology have a dogmatic conception of science that is reflected in the formalism of the analysis, e.g., the theory of manifolds and part-whole logic.

My response: these investigations are not historiographies. They can loosely be described as historically sensitive analyses of persisting philosophical issues in the philosophy of ecology. There are several differences between my phenomenological approach and the historiography tradition. These investigations are not motivated by an interest in the history of science merely as a historical matter, but these genealogies are opportunities in which I clarify and advocate for my own philosophical positions that are themselves a departure from standard approaches in the philosophy of science, e.g., Ernest Nagel's model of the reduction of biology to chemistry and physics. My general methodological claim is that phenomenological resources (theory of manifolds and part-whole logic, in particular) are indispensable in addressing the persistent philosophical issues in the ecological sciences. More specifically, a phenomenological theory of manifolds and part-whole logic can provide a philosophical alternative to the tendency toward uncritical idealization in ecology. This tendency is not a bug in the system of scientific reasoning that could one day be removed through technical achievement, but a feature of the basic assumptions and explanatory principles that demand a continual regard of the things themselves. The ecological things themselves in the North American case studies of the first investigations are Lake Michigan sand dunes, Nebraska prairies, and Cedar Bog Lake of northern Minnesota. These case studies were chosen because of my general familiarity with them and because they are the subject matter of three distinguishable logics of habitat associations in some of the classical studies in early 20th century American plant ecology. Each of these part-whole logics of plant associations makes an important contribution to the accumulation of research in plant ecology, but are not related in my view in any progressive sense. To be sure, there are breakthroughs into new ways of thinking and experimental designs, but they are not related, in my view, to any historical synthesis in which the science of plant ecology realizes itself. The specter of Hegel does not haunt these investigations. My general view is that there are philosophical puzzles that inevitably are part of any modern scientific endeavor and theoretical reflection on them is not only a legitimate enterprise in itself, but can enrich contemporary scientific discourse. The individualism–holism debate in community ecology harbors one of those philosophical puzzles and the early 20th century American plant ecologists are illustrative examples of a plurality of approaches with distinguishable assumptions and explanatory principles.

These investigations are not historiographies. If anything, the genealogical orientation of these investigations is indicative of the view that there is always philosophical spadework to be done in any scientific domain. The individualism–holism debates in community ecology are points of entry into philosophical resources (theories of manifolds and part-whole logic) that can contribute to a propositional understanding of the natural environment in a way that enhances scientific communication and the public discourse concerning environmental issues. I think that phenomenological methodology can contribute to the propositional understanding of how plants are geographically distributed. The genealogical orientation of these investigations is not of historiographies that reconstruct the theories of the past in light of my contemporary viewpoint. They are, generally speaking, investigations about how best to talk about the ecological things themselves without succumbing to the tendency to uncritically idealize them. Do I pick winners and losers in these investigations according to their alignment with my philosophical positions? Yes, I do. I agree with the historians of plant ecology, whether whiggish or not, who distinguish two scientific traditions in late 19th century European plant geography and early 20th century American plant ecology: 1) physiognomic tradition of Humboldt, Oscar Drude, and Clements, and 2) the physiographic tradition of Warming and Cowles. Leaving the historians behind, I critique the physiognomic epistemologies that find philosophical justification in Kant's account of organic form and develop a logic of habitat associations that can complement physiographic epistemologies that operate with a nutritive conception of plant forms. In short, I think that a retrieval of the essential insights of physiography can help preclude the persistent tendency toward uncritical idealization in ecology. I hope that the logic of habitat associations that are developed in these investigations can bolster further physiographic research toward that end. My ultimate aim, however, is a more direct attempt to grapple with the philosophical puzzles involved in the general claim that ecological wholes are more than an aggregate of ecological parts.

A third potential objection to these investigations could be labeled the "semantics objection." According to this objection, these investigations are misguided in their focus of the logical aspects of the individualism–holism debates, rather than a data-driven approach to the cold, hard facts. The line of reasoning in this possible criticism could construe the individualism–holism debates in early 20th century plant ecology as a giant misunderstanding stemming from the (in)accuracy of metaphors. When plant associations are metaphorically referred to as "societies," "communities," or "networked systems," our understanding of these associations is inevitably colored through the linguistic (in)accuracies of metaphors. Viewed in this light, these investigations are at best strenuous labors with the crude tool of metaphors, and at worst a lot of spilled ink over the metaphorical imperfections of past theorists. This objection could be specified further to the differences among classification systems of plant associations of the early American plant ecologists that factor surface groundwater and soil moisture content, on the one hand, and precipitation on the other. In the end, they were both concerned with water variations and their disagreements were largely semantic distractions from continued and careful field research.

My response to this "semantic objection" involves a preliminary clarification of the differences between metaphorical and logical propositions. Metaphors are figures of speech that invoke a comparison of two unlike things that are nevertheless similar in certain aspects. Metaphors are important and necessary modes of propositional understanding and can even help stimulate the investigative imagination of productive researchers. These investigations into the logics of habitat associations and conceptions of plant forms, however, do not rely on metaphors. Identities-in-a-manifold and part-whole relations are not metaphors. Rather, they are logical categories that universally and necessarily apply, in principle, to any object whatsoever. These investigations attempt to reconstruct the basic assumptions and explanatory principles of the classifications of plant associations of the early plant ecologist in an attempt to identify and clarify the basic similarities and differences among them. These logical categories play a different role than metaphors in the propositional understanding of the geographical distribution of plants. I think that a phenomenological approach to these basic logical categories, e.g., the logic of fitness in Husserl's theory of parts and wholes, provides advantages to the standard models that presuppose asymmetrical relations of complex objects with parts and wholes. The classifications of habitat associations and conceptions of plant forms in the history of plant ecology are examples of such complex objects that elicit philosophical reflection in a way that can enhance propositional understanding.

In summary, allow me to attempt to succinctly crystalize the thematic focus of these investigations – the logic of ecological forms. These philosophical investigations are attempts to identify several fundamental assumptions and explanatory principles in the empirical sciences of plant ecology. The contemporary sciences of population and community ecology are empirical sciences that employ multi-level explanations to account for ecological change. Ecological explanations are complex due to the complexity of ecosystem changes. The focus of these philosophical investigations is on ecological explanations concerning plant associations – groups of interspecific plant populations that are characteristic of particular kinds of habitats. The empirical research into the causes of plant population abundance and distribution involve experimental designs and theoretical models that make assumptions regarding the forms of plant associations and their spatial distribution. These investigations concern the suppositions concerning the forms proper to plant associations.

Notes

1 Immanuel Kant, *Kants gesammelte Schriften I, Bd 3, Akademie Ausgabe* (Berlin: de Gruyter, 1902), A 74/B 100.
2 Herbert Butterfield, *The Whig Interpretation of History* (New York: W. W. Norton, 1965), 11.

1 Varieties of succession

A genealogy of early 20th century plant ecology

1.1 Introduction

The three case studies of this investigation are Henry Chandler Cowles' account of Lake Michigan sand dune succession, Frederick Clements' account of Nebraska prairie succession, and Raymond Lindeman's account of the lake succession of Minnesota's Cedar Bog Lake. The habitats in these case studies uniquely illustrate different elements and processes of plant succession. First, the freshwater sand dunes along the southern and eastern shores of Lake Michigan are inhospitable habitats that illustrate the initial or primary stages of succession. Second, Nebraska prairies have a regional scale in which various populations of grasses are distributed in a way that illustrates the role of precipitation and temperature (climate variations) of prairie succession. Third, Cedar Bog Lake is a senescent (late stage) lake that illustrates a culminating stage of inland lake succession in which the distribution of aquatic plants transitions into a bog forest. These case studies are geographically unique habitats that illustrate characteristic changes in the abundance and distribution of plant populations. The accounts of these particular habitats contributed to a more general theory of plant succession in the early 20th century in a way that illustrates different fundamental assumptions and explanatory principles. Generally speaking, these assumptions and principles not only illustrate differences within early 20th century plant ecology, but the broader multi-disciplinary theoretical models at work in the ecological sciences. As Oscar Drude stated, "These considerations show us that ecology is the borderland to which the sciences of biology and geography can both lay claim."[1] This investigation is an exploration of the explanatory borderland at work in these various accounts of succession.

The accounts of succession involved in these case studies operated, more specifically, with different logics of habitat associations. Cowles' account of dune succession, Clements' account of prairie succession, and Lindeman's account of lake succession – each of these contributions to the theory of succession in early 20th century plant ecology have identifiable logics of part-whole relations that are presupposed by their classification of plant associations. Which part-whole relations are involved in the assumptions regarding plant associations in these various accounts of succession? The first case study in this investigation is Cowles' account of freshwater sand dune succession.[2] Cowles' account of the primary succession of Lake Michigan sand dunes operated with a conception of plant association that has a reciprocal dependency with the physiographic (surface geology and topology) features of the habitat. The part-whole logic that is supposed by Cowles' account of succession involved a mutual or reciprocal founding relationship between the topological features of sand dunes and the physiological features of pioneering plant populations, e.g., high rate of vertical shoot growth, rhizomic root structure, and low moisture tolerance. The identification of Cowles' logic of dune habitat associations clarifies the directional attributes at work in his account of succession as oriented toward a physiographic base level, e.g., topological variations involved in erosion and sediment deposition.

The second case study in this genealogy of 20th century American plant ecology is Clements' account of Nebraska prairie succession.[3] Clements' account of prairie succession operated with a logic of habitat associations that attributed functional and directional attributes to the association as a whole. That is, Clements' logic of habitat associations included and emphasized the functional interactions of plant populations as parts of community with biological attributes proper to itself in an analogous way in which organs are functional parts in the activity of individual organisms. Clements' account of prairie succession operated with a conception of plant association that had emergent organic properties that contribute a directionality to the sequence of prairie succession. The basic features of habitats that Clements employed in his classification of plant associations were climate variations – in particular, precipitation and temperature variations. In his account of the short, tall, and mixed grasslands of the Great Basin of North America, Clements conceived of plant succession as a directional sequence that is oriented toward a climax community that achieves stability through an optimal fit with regional climate variations.

The third case study in this genealogy of plant succession is Lindeman's account of the succession of Cedar Bog Lake.[4] Lindeman's logic of habitat associations examined the abundance of plant populations in terms of annual rates in productivity and, in particular, the chemical constituents of nutritive cycles. Lindeman's trophic dynamic account of shallow water (hydrarch) lake succession transformed questions concerning the successive distribution of plants through a thematization of the variations among seasonal nutrient cycling. The study of hydrarch succession in early 20th century limnology involved a shift in thematic focus – the new object of inquiry became the annual nutrient cycle. Lindeman's account of the late stages of lake succession on Cedar Bog Lake shifted the thematic focus from the distribution of plant populations in terms of phenotypic attributes to a thematization of the distribution of calories through an annual trophic cycle. His account of lake succession also illustrates a logic of habitat associations that takes nutrient productivity as the basic variation of succession through an abstraction to the chemical constituents of caloric transfer rates among trophic levels. Lindeman's account of lake succession illustrates how the study of succession in mid-20th century limnology approached questions concerning succession through a logic of habitat associations that was organized by associative caloric fitness.

The case studies in this investigation illustrate that there was a plurality of fundamental assumptions and explanatory principles involved in early 20th century American plant ecology. My reasons for selecting these particular cases not only include the clarity of the philosophical differences in these accounts, but also because they are historically familiar points of entry into broader philosophical issues that persist in the ecological sciences today. Contemporary research is not primarily concerned with the dynamics of succession as a working theoretical interest that explicitly frames its experimental design, but the problem of succession nevertheless provides a useful point of entry into philosophical issues that persist in the background of this empirical inquiry. One of these philosophical issues concerns so-called "unit of analysis" problems or what Frank B. Golley termed the "confusion of levels."[5] The ecological sciences involve multi-disciplinary approaches that attempt to identify the lawful regularities involved in ecological change. Contemporary ecological sciences utilize multi-level explanations. For example, Eugene P. Odum and Gary W. Barratt identified several levels or "units" involved in analyses of ecosystems, e.g., cell, tissue, organ, organ system, organism, population, community, ecosystem, landscape, biome, and ecosphere.[6] Each of these ecological "units" involves overlapping and intertwining levels of explanation. There are several standard methodological approaches that organize the relations among these ecological units and levels of explanation, e.g., bottom-up reductionist relations in which changes that occur at the level of community are explained in terms of changes that occur in the population level, changes in population abundance and distribution are explained in terms of changes proper to individual organisms, and so on. Another standard methodological approach is a top-down holistic model that explains changes that occur at lower levels of analysis in terms of changes of the emergent properties at higher levels of analysis, e.g., the feedback loops involved in forest canopy structures. These alternate standard models in multi-level explanations can generally be characterized as reductionist and holistic models. The three case studies in this investigation provide accounts of succession that utilize distinct and illustrative assumptions and principles that are operative in contemporary reductionism-holism debates concerning unit of analysis problems.

1.2 Freshwater sand dune succession

Sand dunes are unique geological phenomena that are among the most inhospitable habitats on earth. This uniqueness is evident in the consideration of a variety of factors that comprise dune habitats. First, dunes comprise variable and unstable topographic conditions. Often located near windy shores of lakes and oceans, sand blown by the wind creates topographical conditions that are perpetually shifting. The advance or retreat of a dune through sand drifts can outpace the growth of vegetation. Second, an advancing dune buries previous vegetation and creates a soil structure that is not mediated by previous generations of vegetation. The soil content of dunes is not the rich humus of a forest floor that has accumulated by centuries of plant and animal decay. Rather, a shifting dune wipes the topographical slate clean of previous vegetative succession and is deficient in several minerals needed by most plants. Third, dunes have extreme temperature changes. Dunes are exposed to the accumulating wind effects from lakes and oceans that accentuate cold temperatures, on the one hand, and exposed to direct sunlight that dries the upper layers of sand so completely that the grains of sand lie loose and are susceptible to wind transport, on the other. The daily and seasonal temperature variations on sand dunes are conditions that many plant species do not tolerate.

Topographic instability, primitive soil composition, and extreme temperature variability are factors that contribute to the inhospitality of dunes to vegetation and make dunes unique habitats to study early stages of succession. Dune succession is an example of primary succession – a change in species composition over time in a previously uninhabited environment. Like recently formed volcanic lava and exposed rock scraped clean by glaciers, sand dunes do not have an immediate vegetative history. But while dunes share primitive soil composition and extreme temperature variability with these other kinds of primary succession, dune topographic instability that results from wind drifts is a factor that distinguishes the harshness of dune habitats. Shifting sand dunes are thus a kind of beginning or origin of the process of succession.

Sand dunes are found across the globe and mainly arise along lake and ocean coasts and riverbanks. From northern European coasts to the inland sand fields and deserts of Central Asia and Africa, dunes can be identified where vegetation is lacking and the wind sufficiently prevails to transport sand. The vegetation fitted for the loose soil of shifting dunes is invariably characteristic with regard to its root structure and tolerance of low moisture and extreme temperature variation. Grasses with deeply rooted and richly branched rhizomes and bushes with extensively long roots can easily traverse the loosely aggregated sand and have distinctive adaptations that cope with topographical instability. Consider, for example, the *Psamma arenaria* and *Elymus arenarius* found along the coasts of northern Europe. These pioneer grasses can withstand burial by the sand and grow vertically through newly accumulated sand. The elongated sword-like leaves catch wind-blown sand that, over time, is accumulated into hills as grasses populate a shifting dune. The rhizomes of these grasses further stabilize the shifting sands and contribute to the accumulating height of the dune. Indeed, when Danish botanist Eugene Warming studied these dune formations in the 1890s, he noticed that the height of particular dunes could be determined by the growth rate of the particular grass species that occupied the site. Grasses such as *Triticum junceum* and *Honckenya peploides* border the beach and the base of the dune and had a growth rate that produced low dunes along the beach. The growth rate and vertical shoots of *Psamma arenaria* and *Elymus arenarius*, by contrast, can produce high dunes with predominant lee slopes with a "tongue-shaped heap" (a long tongue-shaped accumulation of sand sloping gradually downwards). As Warming stated, "the form of the dune is largely determined by plant-growth" and classified these grasses as "sand binding" plants.[7] This statement is significant in that it illustrates how a biotic variable (growth rate of the vertical shoots of *Psamma arenaria* and

Elymus arenarius) determines an abiotic variable (topographical height of the dune landscape). The early stages of dune succession along the northern European coast, in other words, are an example of a reciprocal relation between plant populations and topography.

The next stage in the vegetative stabilization of dune drifts involves the establishment of shrub populations. Consider, for example, the dunes along the southern shoreline of Lake Michigan that fluctuate by predominant southwesterly and northwesterly winds. The dunes typically edge the beach and the vegetative cover gradually changes as one proceeds inland from the lake. Like the dunes of northern Europe, Lake Michigan dunes are initially stabilized by grasses such as the sand reed (*Ammophila arundinacea*). These pioneer grasses also begin the accumulative process of enriching the soil composition, which is primarily quartz sand but also includes grains of hornblende, magnetite, and coloration from iron oxide. This soil is highly porous and lacks the cohesion associated with maintaining moisture. These highly drainable soil compositions are also typically low in nutrient content and do not rapidly develop rich humus soil due to the rapid oxidation of the organic matter from high temperatures. With the increase in soil fertility from the pioneer grasses such as the sand reed, the soil accumulates enough nutrient content for the growth of shrubs such as the willows (*Salix adenophylla* and *S. glaucophylla*) and sand cherry (*Prunus pumila*).[8] The willows have a particular fitness with the topographical fluctuations of shifting dunes – when a willow shoot is buried in a drift, roots are sent out with a vertical elongation that can outpace the accumulation of sand from the drift. The shoots from these rapidly growing willows can also be uncovered in drift retreat without serious injury. The establishment of these shrubs further contributes to dune stabilization and height accumulation. As Cowles stated in his 1899 field observations, "In short, the species of *Salix* are able to adapt their stems to a root environment, or vice versa, better than any other plants found along the coast. Hence the willows stand abreast of *Ammophila* as dune-formers."[9] In short, the upper dune formations are populated with rapidly growing shrubs such as the willows and sand cherry that further contribute to topographical stability.

Of the few trees that play a significant role in the early stages of dune stabilization along the Lake Michigan shore are the poplars (*Populus monilidera* and *P. Balsamifera*). The former – the cottonwood – have a different propagation strategy than the willows. The long-range dispersal and wind sensitivity of cottonwood seeds tend to lead them to collect and germinate together in the depressions near the water table on the upper beach. These depressions provide partial shielding from the wind, particularly in dunes along the southern coast. Young cottonwood shoots thus tend to grow in dense clusters. These cottonwood clusters are conditioned by the variability involved in the lightness of cottonwood seeds in correlation with topographical and wind variability. The developmental features of the cottonwood that contribute to its propagation concern the growth rate of the young shoots that is rapid enough to outpace the topographical instability of a dune drift. Both the (1) dense clustering and (2) rapid growth of the cottonwood are features of the fitness of these trees to the dunes along the southern shores of Lake Michigan.

Consider, in particular, the dense clustering of cottonwoods as a feature of habitat fitness. Like the *Ammophila* and willows, the cottonwoods successfully cope with topographical instability through rapid growth rates and tolerance of extreme temperature variation. The cottonwood propagates through seed dispersal, rather than the subterranean propagation of the extended rhizomes of the *Ammophila*. The dispersal of cottonwood seeds is distinctive – each seed has a dense tuft of long white hair resembling cotton balls that can parachute long distance wind transfer. Along with these unusually buoyant parachutes, seeds are produced and dispersed in large numbers in early spring and tend to collect in large numbers along topographical depressions, ravines, and stream sides. How is the wind sensitivity of this type of seed dispersal related to the patch density necessary to inhibit a shifting dune? Cottonwood patches accomplish dune stabilization collectively, not individually, through a process of intra-specific symbiosis. As Cowles stated, "The growth of a cottonwood dune, therefore, is of a symbiotic nature, exactly as is the growth of an *Ammophila* dune, in spite of the great difference between the life habits of these two dune-formers."[10] These pioneer species fill a similar role in primary dune succession despite the significant differences among physiology, propagation, and so on. The primary succession of Lake Michigan dunes is commonly accomplished by *Ammophila*, willows, and cottonwoods. Each species utilizes intra-specific symbiosis to stabilize dune drift, accumulate further sand, and subsequently raise the elevation of the population patch. The wind sensitivity of cottonwood seed dispersal, however, distinctively mediates and conditions the symbiotic accomplishment of dune stabilization. This wind sensitivity affords long-range dispersal that tends to collect in topographical depressions. Combined with a relatively synchronized release and large quantities, this type of seed dispersal makes possible the close spatial proximity of a manifold of individuals whose collective growth is necessary to inhibit a shifting dune (see Figure 1.1).

The stabilization involved in the primary dune succession is accomplished by different pioneer species and Cowles correlated these differences with topographical (1) area, (2) shape, and (3) height variability.[11] First, Lake Michigan dunes that were pioneered by *Agropyrum*, *Ammophila* or *Salix* are capable of indefinite areal expansion due to extensive rhizome and root propagation. *Populus* or *Elymus* dunes, by contrast, basically retain the same area and do not expand in that the rate of dune topographical instability outpaces local areal expansion through seed dispersal. Second, the topographical shape of the dunes varies according to pioneering species. The *Agropyrum* dunes are below a meter in altitude and, like the *Ammophila* dunes, are relatively large and low with slopes almost as gentle as the beach. Unlike the *Agropyrum*, however, *Ammophila* dunes are higher due to *Ammophila's* capacity for vertical growth. The *Salix* dunes are also large, but higher and steeper due to even higher vertical growth rate. The *Prunus* dunes are much smaller but relatively high and steep in correlation with low propagation success but higher vertical growth rate. Third, Cowles correlated dune height according to pioneering species. According to the kinds of explanatory co-variation involved in Cowles' account of areal and shape variation, one would expect that dune height variations are determined by the vertical growth rate of the pioneering species. Cowles maintained that

Figure 1.1 Embryonic dune on the beach at South Chicago, formed by the sand reed, *Ammophila arundinacea.* Leeward trail of sand at the left.

Henry Chandler Cowles, "The Ecological Relations of the Vegetation on the Sand Dunes of Lake Michigan (Continued)," *Botanical Gazette* 27, no. 2 (1899): 178.

due to the high growth rate of cottonwoods (the fastest growth rate of all the dune pioneer species), the dunes that were stabilized by cottonwoods would produce the highest and steepest dunes.

Cowles' research on freshwater dune succession demonstrated that biotic features of a habitat can supervene on abiotic topographical habitat features. As Cowles stated, "The distribution of certain topographical forms coincides with the distribution of the dune formers.... This is only one of many cases where ecology helps interpret physiography."[12] The order of primary dune succession demonstrates a reciprocal relation between the variation of local distribution of specific populations and topographical variation. Moreover, Cowles' account of primary dune succession employed a classification of plant associations with a particular logic of parts and wholes that expressed philosophical assumptions and methodological principles through which he navigated the explanatory "border-land between the sciences of biology and geography."

1.3 The logic of plant associations in Cowles' account of dune succession

Cowles' physiographic account of primary succession on Lake Michigan sand dunes presupposes a particular logic of plant associations. His extension of a physiological conception of plants and emphasis on topological features of terrestrial habitats to questions concerning dune succession illustrates several fundamental assumptions regarding the conditional relations of dune habitats. Historically speaking, Cowles understood his account of dune succession to be an extension of Andreas Franz Wilhelm Schimper's emphasis on local "edaphic" soil compositions and Warming's "physiographic ecology."[13] Cowles' account of dune succession especially appropriated and applied a physiological conception of plant nutrition from Warming and an emphasis on the topographical variations of local edaphic habitats from Schimper. What are the fundamental assumptions and methodological principles that are involved in Cowles' account of dune succession? How was Cowles' physiographic approach reflected in his account of the distribution of plant associations on sand dunes?

Cowles' account of Lake Michigan dune succession involved the development of physiographic (physiological and geographic) assumptions and methodology in several senses. First, his classification of plant associations was based on a nutritive conception of plants that emphasized the role of water in nutritive processes. The succession of plant associations was classified according to a specific feature of dune habitats – the moisture content of the soil. The prioritization of soil moisture variations was informed by the 19th century discovery that plants absorb essential mineral nutrients as water ions. Historically speaking, 19th century advances in plant physiology (e.g., Julius von Sachs' water culture experiments and the new cellular logic of plant cell growth) were applied to questions in plant geography concerning the geographical distribution of plants in the work of Warming and Schimper. Cowles appropriated Warming's classification of plant associations that prioritized water variations in the soil as the distinguishing feature among plant associations. Warming stated, "The nature of the nutrient substratum or edaphic conditions, largely determine the habitats of plants and their topographical distribution; and among all characteristics of the substratum the most important is the amount of water contained."[14] This classification emphasis on water variation was directly informed by the role of water in plant nutrient production: "Lack of water in the soil causes the plants to be ill-nourished, because the roots can obtain nutriment from such soil only with difficulty."[15] Cowles explicitly recognized this hydrological basis in Warming's classification of plant associations, explicitly stating that Warming "made variations in the water content of the soil a basis of classification"[16] and appropriated this classification of plant distribution involved in dune succession through the distinctions between xerophyte (plants that tolerate little soil moisture), mesophyte (plants that tolerate wet or dry soil), and hydrophyte (plants that tolerate wet and aquatic habitats). Cowles classified the plant associations involved in primary dune succession as xerophytic plants with characteristically elongated roots and rhizomes – the plant associations involved in dune succession more generally transition from xerophytic pioneer associations of grasses and shrubs to mesophytic forest associations. In short, Cowles' account of dune succession employed a classification of plant associations that was based on the water variations in the soil and this was an extension of a nutritive conception of plants that emphasized the role of water in plant growth. Cowles was among the first to introduce this physiographic approach in American plant ecology. The essential point is that, following and extending Warming's approach, Cowles' classification of plant associations thematized soil moisture tolerance as the distinguishing characteristic among plant associations.

Second, Cowles' account of dune succession involved a spatial (local, regional, and global) distinction that had implications for his categorization of plant associations. He made a scale distinction between regional and local features of a habitat that condition the distribution of plant associations. These regional factors were the temperature and atmospheric variations that are dependent, in turn, on the earth's latitude and rotation around the sun. Following Schimper, Cowles termed the regional distribution of plant associations "climate formations" and maintained that climate variations (temperature and atmospheric moisture) were the primary determining variable at this regional scale. By contrast, the primary determining variable at a local scale was soil moisture content which, in turn, was primarily determined by variations in surface geology and topology (erosion and deposition) or what Schimper termed "edaphic factors."[17] Cowles made a similar distinction between Drude's regional "climatic formations" (e.g., tropical evergreen forest, deserts in continental interiors, prairies, deciduous forests, and artic tundras), and local or edaphic factors (e.g., soil composition such as moisture, topology, air, and temperature, light exposure, and so on). As Cowles stated, "We may say then that atmospheric or climatic factors determine distribution in the large, while local differences are produced by changes in the edaphic or soil factors."[18] This scale distinction clarifies that Warming and Cowles' physiographic approach to the distribution of plants took local topographical variations as its principle object of investigation.

Not only did Cowles categorize plant associations according to water variations, he also developed a physiographic account of succession that prioritized the topological variations involved in water drainage as the distinguishing condition among plant associations.[19] The logic of plant associations in Cowles' "physiographic ecology" is reflected not only in his classification of plant associations according to water variations in the soil (e.g., xerophyte, mesophyte, hydrophyte) but according to the geographical features of the habitats – surface geology and topography, in particular. Surface geology concerns the original composition of the soil, e.g., rock, sand, clay, marl, and so on. The topographical variations are primarily determined by erosion processes involved in water drainage. Water variations condition topographical changes involved in sediment deposition and accumulation. As Cowles stated,

> The processes of erosion ultimately cause the wearing down of hills and the filling up of the hollows. These two processes, denudation and deposition, working in harmony produce planation; the inequalities are brought down to a base level. The chief agent in all these activities is water, and no fact is better established than the gradual eating back of the rivers into the land and the wearing away of coast lines.... As a consequence of all these changes, the slopes and soils must change; so too the plant societies, which are replaced in turn by others that are adapted to the new conditions. There must be, then, an

order of succession of plant societies, just as there is an order of succession of topographic forms in the changing landscape.[20]

Cowles classified the plant associations involved in dune succession according to the physiographic (surface geological and topographical) features that characterize successive stages.[21] The "normal primitive formation" was the beach – the baseline for the inland sequence from stationary beach dunes, the active or wandering dunes, the transitional dunes, and passive or established dunes.[22] The established dunes pass through several forest stages and eventually culminate in deciduous mesophytic forests. Cowles' classification of the characteristic stages of dune succession was topological – the developmental order of dune habitats concerns the successive variation of topology through the dune's post-glacial history. The directionality that orders these topographical variations into a series with characteristic stages is the geological past to the present – the water drainage of the Wisconsin glacier, in particular. Topological succession is linear in the general sense of the long-term (geological) recession of Lake Michigan.

Third, Cowles' extension of physiology to questions concerning plant distribution involved a particular conception of the unity proper to plant associations. Cowles defined a "plant society" as "a group of plants living together in a common habitat and subjected to similar life conditions."[23] Accompanying topographical succession was an order of succession of plant societies – local interspecific associations of plants that are characteristic of freshwater dune habitats. Cowles appropriated the category of plant society from Warming, stating, "The term [plant society] is taken to be the English equivalent of Warming's *Plantesamfund*, translated into the German as *Pflanzenverein*."[24] Both Cowles and Warming had an individualistic conception of plant associations. Warming's account of plant associations particularly illuminates this individualistic conception of plant associations.

Warming generally characterized *Plantesamfund* (plant societies or communities) as follows: "Certain species group themselves into natural associations, that is to say, into communities which we meet with more or less frequently, and which exhibit the same combination of growth-forms and the same facies."[25] More specifically, Warming defined a plant society or community through a logic that took the individual plant as the primary unit of analysis:

> The term community implies a diversity but at the same time a certain organized uniformity in the units. The units are the many individual plants that occur in every community, whether this be a beech forest, a meadow, or a heath. Uniformity is established when certain atmospheric, terrestrial, and any of the other factors…are co-operating, and appears either because a certain, defined economy makes its impress on the community as whole, or because a number of different growth-forms are combined to form a single aggregate which has a definite and constant guise.[26]

Warming's definition is individualistic insofar as it takes individual plants to be the primary unit of analysis (rather than populations or communities). Plant societies or communities are, first and foremost, aggregations of individuals. Cowles agreed with Warming on this individualistic conception of plant associations, as he metaphorically and succinctly put it, "egoism reigns supreme" on dune habitats. But Warming and Cowles also had a conception of plant associations that operated with a logic of reciprocal dependence through which a plant society accomplishes something collectively through its large-scale organization, e.g., dune stabilization. As Warming stated,

> In [*Plantesamfund*] there is, it is true, often (or always) a certain natural dependence or reciprocal influence of many species upon one another; they give rise to definite organized units of a higher order; but there is no thorough organized division of labour such as is met with in human and animal communities, where certain individuals or groups of individuals work as organs, in the wide sense of the term, for the benefit of the whole community.[27]

Plant societies do not have units that function in light of a benefit for the whole, nor are they integrated in the way that individual plants are integrated with distinct organs that fulfill a distinct nutritive function. The geographical distribution of plants nevertheless displays "definite organized units of a higher order." The collective attributes of plant associations – the "societal" features – are of a "higher order" insofar as the various symbiotic and competitive interactions among individuals are not reducible to individual achievements. The symbiotic accomplishment of dune stabilization illustrates how the growth of interspecific populations as a whole achieves a fit that is not and cannot be achieved by individual plants. In short, Warming and Cowles' account of the unity of plant societies was individualistic – the reciprocal dependence of organic parts is proper to the individual organism and did not generalize to plant collectives such as populations and communities. The holistic accomplishments of plant associations are not primarily the result of an organic interdependence of individuals, but an aggregated large-scale organization that has a reciprocal dependence on topographical variations.

Fourth, Cowles' account of succession in general involved a directionality that was oriented to a topographical base level. Not only do water variations organize the nutritive process, but these water variations also organize the topographical variations that condition plant distribution. In other words, the same water variations that condition plant distribution also condition the water variations involved in non-vegetative topographical changes, e.g., erosion and sediment accumulation in swamp recession. Consider, for example, the role of water in the topographical changes involved in erosion. The process of erosion wears down hills (denudation) and fills up hollows (deposition) to produce an overall tendency to a base level. While the original soil composition is determined by surface geology, e.g., rock, sand, and clay, the topographical changes involved in erosion are particularly determinative of plant distribution in that they condition drainage, air content, and humus. Consider, for example, the lack of drainage, low aeration, and sediment accumulation in peat bogs – these topographical changes are features of the same water variations that condition plant distributions into hydrophytic and mesophytic associations. Water variations are the identity in both manifolds of topographical changes and the nutritive processes involved in plant growth. This identity provides a notion of necessity in Cowles' account of succession that is lost in theories that correlate two distinguishable variables. The water variations that determine the topographical changes in processes of erosion have an overall base leveling effect – hills are worn away and hollows are filled – and provide a directionality to these topographical changes. These changes are successively ordered toward a base level – large flood plains with mesophytic communities found, for example, in the lower Mississippi River valley. In the relatively young topography of recently glaciated regions of Michigan, Wisconsin, and Minnesota, by contrast, there is a broad variety of topographical conditions and plant distributions, e.g., hydrophytic lakes and swamps and xerophytic hills. In passing from earlier to later stages in topographical succession, a region loses its topographical variation, and, with it, the plant distributions involved in its hydrophytic and xerophytic hills. As Cowles stated,

> the ultimate stage of a region is mesophytic. The various plant societies pass in a series of successive types from their original condition to the mesophytic forest, which may be regarded as the climax or culminating type…through all these changes and counterchanges the great mesophytic tendency is clearly seen; mesophytic areas may be lost here and there but many more are gained, so that the approach to the mesophytic base level is unmistakable.[28]

The directionality involved in Warming and Cowles' account of plant distribution is not derived from an organic notion of function that is overgeneralized to the holistic features of plant collectives. Rather, this directionality is derived from the tendency of topographical base-leveling involved in erosion and sediment accumulation.

Fifth, Cowles' account of plant associations was ideographic. The surface geological and topographical variations that organize the distribution of plant associations are historical – they are understood through the geological past of the habitat. As he stated, "to many of us it has seemed that the first step is the study of the life histories of these [plant] associations in each region."[29] This introduction of geological history into the account of plant distribution implies that the explanation is ideographic – it explains how the topography has changed but does not demonstrate that these changes could not have happened otherwise. The explanatory principles in Cowles' "physiographic ecology" were descriptive patterns concerning the geographical distribution of plants, not a mechanistic explanation that is universally lawful without exception. This is the sense in which Cowles characterized his account of dune succession as "a variable approaching a variable rather than a constant."[30] Cowles' account of succession operates with two related manifolds of variation – topographical and nutritive. The common feature that is determinative of both manifolds is water variations. The same, identical water variations are the determining condition for the manifolds of topographic and nutritive change. Cowles understood the explanatory purchase of his account of dune succession to describe particular local differences, stating, "the orders of succession are not the same in various regions…each region must be worked out for itself."[31]

Cowles' investigations of freshwater sand dune habitats uniquely illustrate the dynamics involved in primary succession. The wandering sand dunes of Lake Michigan are among the most inhospitable habitats on the planet due to the high rate of topological instability, extreme temperature range, and low nutrient content and low moisture of the soil (quartz sand). Almost nothing grows there. This absence of vegetation uniquely illustrates a genetic origin of the developmental features of the distribution of plants. It is in this sense that Cowles metaphorically characterized a wandering dune that has been initially stabilized by populations of pioneer dune grasses as a "dune embryo." As he stated, "The primary motive, then, which prompted this present study was the feeling that nowhere else could many of the living problems of ecology be solved more clearly; that nowhere else could ecological principles be subjected to a more rigid test."[32] This account of dune succession illustrates a methodological intertwinement between physiography and ecology. In particular, his account of plant succession was founded upon the topographic succession involved in water drainage – explanations of the distribution of plant associations that presuppose and form a unity with explanations of topographical variations. Succession as an explanatory principle that addresses the questions concerning plant distribution is logically dependent on the explanatory principles in physiography. Moreover, Cowles' account of Lake Michigan dune succession illustrates how this logical dependency is reciprocal – dune topographical variations presuppose and form a unity with the dune forming grasses and shrubs. In this sense, ecology helps interpret geography. These reciprocal founding-founded relations are made possible through the thematic focus on water variations as the identity in both (topographical and nutritive) manifolds of change. Thus, there are reciprocal dependencies between topographical changes and plant distributions in the primary stages of succession.

The logic of habitat associations in Cowles' account of succession was different than that of Clements'. This difference grounded the debates between the so-called Chicago and Nebraska schools in the 1910s and 1920s. These debates have been documented by historians. My investigation in this chapter is not primarily concerned with the sociological history of plant ecology. This investigation is different from these previous historical treatments in its focus on the difference between the fundamental assumptions and explanatory principles at work in the respective accounts of succession. This focus is not primarily a historical concern, but a description of case studies that manifest philosophical differences that are persistent in broader ecological discourses. As Cowles stated,

> Recognizing, then, the interdependence if not the complete identity of experimental ecology and field physiology, it becomes necessary to consider the

underlying philosophy that serves as a motive for all research along these lines. The importance of one's philosophy upon his research can scarcely be overestimated, for it determines the problems to be attacked, the methods to be employed, and, on account of the personal equation, is likely to give some color to the results secured.[33]

In short, it is possible to reconstruct the basic assumptions and explanatory principles in Cowles' account of dune succession through an investigation of his classifications of plant associations. This identification of the logic of habitat associations can be clarified through a contrast with Clements' account of prairie succession.

1.4 Nebraska prairie succession

Meriwether Lewis and William Clark's expedition to discover a northwest passage across North America began along the confluence of the Missouri and Mississippi Rivers at Camp Dubois in 1803. They travelled through the deciduous forests that extended into the Missouri River Basin, through the grasslands of the continental interior, and to the northwest conifer forests before reaching the Pacific coast. The oak-hickory forests beyond the Appalachian Mountains had not yet been timbered and the prairies west of the Mississippi River had yet to be plowed. In 1809, John Bradley surveyed the grasslands of the North American interior and described an abundance of grass populations capable of sustaining extensive herds of large mammals such as bison, elk, antelope, and deer. The regional extent of these grasslands proceeded west of the Mississippi River almost to the Rocky Mountains and from the conifer forests in the north to the tundra and deserts to the south.[34] Along with the distribution of these forests and tundra, the grasslands of the continental interior were so vast that they warranted distinct regional categorization. When Clements and Roscoe Pound published *The Phytogeography of Nebraska* in 1897, the vast distributions of prairie associations were still uncultivated by the plow. The distribution of prairie associations in the Great Plains of North America prior to 20th century agricultural development was not only distinctive in the vastness of its regional scale, but its long-term persistence. In contrast to the more locally scaled patchwork of fields in Europe that had been cultivated for centuries, the open grasslands of the North American continental interior at the turn of the century afforded Clements and Pound prairie habitats that were not significantly altered by agricultural development.

The surface geology of the Great Plains region is primarily influenced by the last glacial advance and retreat during the Pleistocene Era – an ice age that dates back about 2.6 million years ago. The last glacial retreat was approximately 18,000 years ago, one that uncovered a landscape denuded of vegetation and soil. The topographical depositions were largely outwash plains that ultimately drained into the Mississippi River east of the continental divide.

The climate variations across the Great Plains involve a gradual eastward increase in mean annual precipitation and decrease in mean annual temperatures.

Clements summarized his pioneering data regarding the regional precipitation and temperature as follows,

> In terms of precipitation, the latter may range along the parallel of 40° from nearly 40 in. at the eastern edge of the true prairie to approximately 10 in. at the western border of the mixed grassland, or even to 6 in. in the desert plains and the Great Valley of California. Such a change is roughly 1 in. for 50 miles and is regionally all but imperceptible. The temperature change along the 100th meridian from the mixed prairie in Texas to that of Manitoba and Saskatchewan is even more striking, since only one association is concerned. At the south the average period without killing frost is about 9 months, but at the north it is less than 3, while the mean annual temperatures are 70 and 33°F. respectively. The variation of the two major factors at the extremes of the climatic cycle is likewise great, the maximum rainfall not infrequently amounting to three to four times that of the minimum.[35]

The mean annual precipitation across the 40° parallel increased west to east – the eastern prairie associations received four times as much precipitation as the middle prairie associations and almost seven times as much as the desert plains and westward into California. The mean annual temperature increased north to south along the 100th meridian. From Texas to the south and Manitoba and Saskatchewan to the north, there was a decrease in mean temperatures and length of growing season. In short, the eastern prairies were relatively moist, the desert plains were relatively dry, and the California prairie had winter rainfall whereas the Palouse prairie had winter snowfall.[36] These precipitation and temperature variations were the fundamental feature of grassland habitats in Clements' classification of plant associations and account of prairie succession.

Clements classified prairie associations as a type of "grass form" in contrast to tree and shrub forms. Plant associations of the tree form genus have three "sub-forms" – coniferous, deciduous, and broad-leaf evergreen and are typified in boreal, temperate, and tropical forest associations. The three shrub associations that are characteristic in North America are desert, sagebrush, and chaparral. Two characteristic grass forms can be distinguished – the prairie and steppe (semi-desert populated with grasses and shrubs).[37] Generally speaking, prairie associations blend and compete with shrub associations to the west and south and forests to the east and north. Clements classified six prairie associations: (1) mixed grass, (2) true prairie, (3) tall grass (*Andropogon*) prairie, (4) desert plains, (5) California prairie, and (6) Palouse prairie.[38] These prairie associations were characteristically distributed regionally – the mixed prairie (composed of mid-grass and short-grass populations) is geographically central to the others. To the east along the Missouri and Mississippi Rivers, the mixed prairie transitions eastward into the true prairie with mid-grass populations and transitions into tall-grass (*Andropogon*) associations that, in turn, are bordered by deciduous forests. To the south, the coastal prairies extended into the Gulf regions of Texas and Mexico and the desert plains (primarily characterized by species of *Bouteloua* and *Aristida*) ranged from western Texas before transitioning into the desert plains of Arizona and Mexico. To the northwest, the short grasses dissipated into the Palouse prairie association of eastern Washington.

The distinguishing feature of these prairie associations in Clements' classification system was the abundance of dominant species. Clements identified *Stipa spartea*, *Sporobdus asper*, and *heterolepis* as abundant in true prairie associations, *Bouteloua eriopoda*, *rothrock*, and *radicosa* and *Aristida california* as abundant in desert plains, and *Stipa comate* and *Buckloe* in mixed prairie associations. Mixed prairie associations were the most extensive and varied. Not only were mixed prairies centrally located in geographical relation to the other five prairie associations, but it included species from them. *Stipa*, *Agropyrum*, and *Koeleria* transitioned to the east and *Holarctica sporobolus* migrated to the south, and the short grasses from the Mexican plateau. The mixed prairie association included many of the genera that became dominant in the other prairie associations and Clements viewed it as ancestral to them, that is, he viewed the centrally located mixed prairie associations as a point of departure for the distribution of the other associations characterized by various dominant species that are more tolerant of regional habitat variations (especially precipitation and temperature variations). In contrast to the conventional view at the time which maintained that forests would eventually succeed the grasslands, Clements and Pound discovered that grasslands were actually expanding geographically.[39]

Clements' account of the regional distribution of prairie associations did not merely include static descriptions of the spatial extent of dominant grass populations, but emphasized the dynamic development of associations through a process of prairie succession. Generally speaking, Clements' account of prairie succession provided a developmental explanation for the regional distribution of prairie associations. As indicated earlier, the distribution of prairie associations in Clements' account was primarily conditioned by the variations in precipitation and temperature across the continental interior – grass populations were abundant depending on the considerable climate variation across the region. While a static classification merely correlates the variation of population abundance with these climate variations, Clements' dynamic account of prairie succession appealed to developmental features proper to prairie associations as a whole and provided functional explanations regarding grass distributions. Historically speaking, Clements appropriated Drude's notion of "plant formation" to characterize the developmental features of prairie associations as communities with organic attributes proper to its collective interactions. Clements clearly stated this developmental and holistic conception of succession in the preface to *Plant Succession* (1916):

> The essential nature of succession is indicated by its name. It is a series of invasions, a sequence of plant communities marked by the change from lower to higher life-forms. The essence of succession lies in the interaction of three factors, namely, habitat, life-forms, and species, in the progressive development of a formation. In this development, habitat and population act and react

upon each other, alternating as cause and effect until a state of equilibrium is reached. The factors of the habitat are the causes of the responses or functions of the community, and these are the causes of growth and development, and hence of structure, essential as in the individual. Succession must then be regarded as the development or life history of the climax formation. It is the basic organic process of vegetation, which results in the adult or final form of this complex organism. All the stages which precede the climax are stages of growth. They have the same essential relation to the final stable structure of the organism that seedling and growing plant have to the adult individual. Moreover, just as the adult plant repeats its development, i.e., reproduces itself, whenever conditions permit, so also does the climax formation. The parallel may be extended much further. The flowering plant may repeat itself completely, may undergo primary reproduction from an initial embryonic cell, or the reproduction may be secondary or partial from a shoot. In like fashion, a climax formation may repeat every one of its essential stages of growth in a primary area, or it may reproduce itself only in its later stages, as in secondary areas. In short, the process of organic development is essentially alike for the individual and the community. The correspondence is obvious when the necessary difference in the complexity of the two organisms is recognized.[40]

This general conception of plant succession can be illustrated through prairie succession in particular. A prairie association is a collection of interspecific grass populations that function as organic parts in relation to the prairie community. These grass populations have functional responses to features of the habitat, especially precipitation and temperature variations, that condition their abundance and distribution. The developmental sequence involved in prairie succession culminates in a climax prairie community that has reached a stable equilibrium with its habitat. The sequence is composed of a series of prairie associations (serial formations) that gradually culminate in a formation that achieves a stable equilibrium (optimal fitness) with regional climate variations.

Clements' account of prairie succession was informed by field data that employed the meter plot or quadrat. This method of measurement involved marking out a square area of various sizes, using a meter tape to plot out a Cartesian coordinate grid, and counting the grasses within the plot along with their coordinate locations. Clements developed five types of quadrat measurements: list, chart, permanent, denuded, and aquatic quadrats. First, the list quadrat recorded the total number of individuals in the various plant populations present in the quadrat. The abundance of these grass populations was presented in a list. Second, the chart quadrat spatially represented the distribution of grass populations from the list quadrat through plotting the individuals along Cartesian coordinates. Third, the permanent quadrat involved longitudinal measurements over annual seasons in an attempt to identify grass populations that persisted throughout distributional variations. Repeating the data collection of these quadrats over several years, Clements enumerated the variations of abundance and diversity of grass populations in a given location over time. This quadrat methodology not only revealed characteristic prairie associations (serial formations) replacing others in later successive stages, but could also be modified to investigate earlier stages of prairie succession. Fourth, Clements devised what he termed a "denuded" quadrat that involved removing all the vegetation from the plot and recording its regrowth over several years. In such cases, it was possible to record the abundance and diversity of pioneering grasses in the early stages of prairie succession.[41] In short, the mathematization of Nebraska prairies through the quadrat was a methodological advance that clarified various characteristic serial associations of grass populations. Historically speaking, Clements' analysis of Nebraska prairies according to this quadrat methodology represented an advance from a floristic or physiognomic study of plant associations and provided quantitative field data for measuring the serial stages involved in prairie succession.

1.5 The logic of plant associations in Clements' account of prairie succession

What is the logic of habitat associations that is involved in Clements' account of the prairie succession? Clements defined succession as "the universal process of formation development" that includes "succession as the growth or development and the reproduction of a complex organism. In this larger aspect succession includes both the ontogeny and the phylogeny of climax formations."[42] The term "ontogeny" refers to the developmental stages of the lifespan of organisms. By contrast, the term "phylogeny" refers to the evolutionary relationships among various biological species. The central explanatory principle of Clements' account of prairie succession concerned the developmental or dynamic stages of plant associations as organized wholes that are analogous to the developmental stages of individual organisms. The collective itself has developmental stages that have organic properties. The distribution of plant associations can be explained through the phylogeny that is proper to plant associations as organized collections (wholes). Clements thus also extended the notion of evolutionary relationships of biological species to the evolutionary relationships among plant associations. The logic of plant associations in Clements' account of prairie succession included the following features:

1) The stages of prairie succession are comprised of the variation of serial formations.
2) A sere (serial formation) is the basic unit of succession and a local distribution of characteristic grass associations.
3) A serial formation is a developmental unit – it progresses from early serial formations composed of pioneer grass populations to a climax formation. A sere is a "vegetative formation" in Drude's sense of the term.
4) A "cosere" (literally, "binding together in a whole") is the spatial convergence of similar or related serial formations into a developmental bond of the serial formations.

5) A cosere (developmental bond serial formations) progresses from instability to stability through the internal regulation of climate variations, e.g., the regulation of temperature through canopy structures, moisture retention of humus, and so on.

6) The development of serial formations can be functionally described through the following progressive series: migration of pioneer species, ecesis, competition, reaction, stabilization.

7) Under optimal climatic conditions, the serial formations culminate in a climax community.

8) A climax community is phytogenetic in the sense that its biological properties have remote causes that explain its life-history and proximate causes of the functional features proper to the integration of the whole.[43]

An initial point of entry into these features of Clements' account of prairie succession concerns the holistic features of prairie associations as collective units. As Clements stated, "The developmental study of vegetation necessarily rests upon the assumption that the unit or climax formation is an organic entity."[44] For Clements, the prairie associations involved in climax formations have organic properties proper to themselves as a whole that are analogous to the organic properties of individual plants. As he continued, "The life-history of a formation is a complex but definite process, comparable in its chief features with the life-history of an individual plant."[45] Clements made this analogy even more explicit when he stated, "In short, the process of organic development is essentially alike for the individual and the community."[46] The development that operates at the community level is inherently and inevitably progressive, which is to say, the developmental features at the community level progress from germination to maturation in a unidirectional series that plateaus in a stable equilibrium of the climax community, "while the course of development may be interrupted or deflected... whenever movement does occur it is always in the direction of the climax."[47] In Clements' view, in short, a prairie association is a complex vegetative unit with a developmental process that is typified by its climax association. This account of prairie succession introduces a certain kind of internal regulation that illustrates his conception of a prairie association as a vegetative unit with a developmental process proper to itself as a whole. As various grass populations compete for light exposure, soil nutrients and moisture, and root structure space, the populations that are more tolerant of habitat variations eventually increase in abundance through interspecific symbiosis that increasingly stabilizes the soil composition of the prairie.

Second, the classification of climax communities in Clements' account of prairie succession was organized by the dominant grass populations – a prairie association develops toward a climax association that is most fitted to a particular climatic region. In contrast to the topological variations that organized Cowles' physiographic classification of plant associations, precipitation and temperature variations were the habitat features that were given the prioritized explanatory role in Clements' classification of climax formations. More specifically, Clements characterized a physiographic account of succession as concerned with "remote causes" that are involved in initial stages of succession in contrast to the "proximate causes" involved in temperature and precipitation variations. Whereas physiography can adequately explain the remote causes of primary succession, in Clements' view, he did not think it could adequately explain the proximate causes of the development of plant formations as holistic vegetative units as they are determined by temperature and precipitation variations. As Clements stated, "The influence of physiography in this respect is controlled or limited by the climate.... These are subordinate as causes to the general terrestrial climates, which are the outcome of the astronomical relations between the sun and the earth."[48] Not only do climate variations provide a clearer analysis of the later successive stages of regional plant formations, but they address the "chresard to which the plant responds, and not the soil-texture or the physiography."[49] The "chresard" (soil moisture content) is primarily conditioned by precipitation and temperature variations, rather than the remote variations conditioned surface geology and topology.

Third, Clements' distinction between sere and cosere appropriated a logic of plant associations from Drude's notion of "vegetative formation."[50] It is worth examining conception of form more closely in that it illustrates an epistemological idealism that is also reflected in Clements' account of prairie succession. Drude provided a systematic articulation of what he called "the modern doctrine of 'plant formations'" in his address entitled "The Position of Ecology in Modern Science" at the Congress of Arts and Science meeting in St. Louis, Missouri in 1906.[51] This address succinctly articulates the basic theoretical questions of ecology, the historical development of the current notion of "plant formation," and highlights Drude's own articulation of this theoretical concept. First, Drude's address was written during the early decades of the 20th century, the time in which plant ecology first became self-conscious of itself as an interdisciplinary scientific discipline with a specific and unique subject matter. He generally characterized the discipline of ecology as "the borderland to which the sciences of biology and geography can both lay claim"[52] and defined ecology as follows:

> The ecological point of view includes those things concerning the question of continuance in a given location, the power of obtaining nourishment, and the certainty of establishing the succession, which are not general and uniform, but which differ according to the varying factors of the environment. Ecology is the study of the epharmony in the organic world, and the possibility of variation possessed by species, as well as the mere fact of their life together, is an ever-present, dynamical factor in the determination of the external appearance of our earth. But it is not enough simply to discover and

describe these various specific relations; we must press forward and from the mass of accumulated data obtain an intimate comprehension of the organic form in its dependence upon Mother Earth![53]

This definition of ecology is noteworthy for its attempt to be comprehensive of both plant and animal ecology, the characterization of formations as organic in general, and the explanatory difference between description and comprehension. Also particularly noteworthy is the use of the term "world" to describe the totality of living phenomena – an important analytic tool in the Post-Kantian tradition to characterize the universality of various kinds of unity that are characteristic of various kinds of theoretical interests. For Drude, the ecologist's interests in questions concerning plant distribution, succession, and physiography involved the formulation of a regional vegetative unit of analysis having emergent properties that causally conditioned the development of organic (plant and animal) individuals and groups.

Second, Drude's definition of ecology synthesized various historical contributions to the notion of plant formation into five historical periods that, together, comprise the modern doctrine of plant formations. In particular, while Drude credits August Grisebach with thematizing the climatic conditions of plant formations, Grisebach lacked the necessary accumulation of data to comprehend the intimate relations between climate and plant life.[54] Schimper's contributions advanced Grisebach's project and established the classifications and explanations for the climatic limitations involved in plant formations. Drude presents Schimper's contribution in contrast to Warming's physiography, which focused on small-scale micro-phenomena with the fine-grained physiological analysis, and as "completely restoring" the connection between large-scale geographical phenomenon and climate variations. While Warming's physiographic approach at first seemed to usurp Grisebach's conception of vegetative form, Schimper's work re-established the methodological orientation toward questions of plant distribution and succession through the use of a plant formation as a unit of analysis and explanation.

Third, Drude presented a systematic articulation of the notion of formation through three inseparable points of view. An "ecological formation" (one that includes both plant and animal formations) is comprised of three basic relations:

1) "the relation of the organization of ecological forms to the morphology of plants and animals (morphological relation),
2) the relation of the ecological formation to the physiography of the country (physiographical relation),
3) and the relation of the ecological epharmony to the phylogeny of systematic groups in both animal and plant kingdoms (phylogenetic relation)."[55]

First, an ecological formation organizes and is organized by the morphology (anatomy and physiology) of the organism, e.g., the leaf structure of deciduous trees, which "bears the imprint of the climatic seal and determines the fundamental form" of the organism.[56] The morphological relation provides classification of different ecological formations such as forest, prairie, dune, swamp, and so on. Second, an ecological formation consists in a geographical relation that determines the distribution of plants and animals – both in the small-scale sense of physiography and in the larger regional sense. Third, an ecological formation is comprised of a relation to phylogeny, which Drude defines as "the study of the variability of species during the struggle for space under the direct influence of newly acquired qualities."[57]

This third phylogenetic relation represents Drude's distinctive contribution to the notion of regional vegetative form. While Drude's articulation of the phylogeny of ecological formations did not persist in subsequent generations of plant ecology, it significantly influenced how Clements characterized regional plant formations with biological properties proper to itself, e.g., Clements' conception of a climax community such as prairie association as a complex super-organism. This significance can initially be indicated by Drude's phrase "struggle for space" as a way to spatially represent the process of evolution. Like Warming, Drude saw the implications of a theory of evolution for questions concerning plant distribution. But while Warming conceived of evolution by natural selection as occurring in local topographies fundamentally conditioned by water variations, Drude applied an evolutionary approach to plant distribution in regional formations fundamentally conditioned by climate variations. This application, in Drude's view, proceeded along two lines,

> according to the variation of species in regard to their spatial requirements, or according to the variation of an association under the influence of successive generations, each of which has undergone modifications. In this way the study of phylogeny is extended to the field of floristic geography.[58]

The first line of evolutionary thought has a Lamarckian emphasis on the direct influence of climate variations on species composition, e.g., Gaston Bonnier's *Géographie botanique expérimentale*, and the second line of evolutionary thought is more Darwinian in its emphasis on long-term adaptation through interspecific competition. However, this second line of evolutionary thought regards the variations proper to the association, not the individual species that comprise the association. Neither of these two lines of evolutionary thought are specifically Darwinian in their emphasis on the adaptation of species through a long-term process of natural selection. Drude's articulation of the phytogenic relations of ecological formations, rather, was Lamarckian in its emphasis on the direct influence of the environment on the short-term adaptation of species. Drude extended this Lamarckian interpretation of evolution to apply to plant associations.

The two lines of evolutionary thought that are involved in Drude's articulation of phylogenetic relation of ecological formations prefigure Clements' distinction between proximate and remote causes. Recall Clements' characterization of Warming's physiographic account of succession as concerned with "remote causes" that are involved in initial stages of succession in contrast to the "proximate causes" involved in temperature and precipitation variations. Clements' account of plant succession attempted to causally explain the proximate causes of the development of plant formations as holistic vegetative units as they are determined by regional climate variations. In particular, regional precipitation variations primarily determine the "chresard" (soil moisture content) to which plants respond. The Nebraska prairies are flatlands with minimal disruptions to their surface geology and their late-stage Mississippi River watershed did not illustrate the physiographical variations that are evident in more recently glaciated regions. Water variations in the soil help explain the distribution of prairie associations, but in Nebraska prairie habitats, soil moisture variations are primarily determined by precipitation variations (rather than physiographic features such as erosion and deposition). As Clements stated, climate variations are the "chresard" to which the plant directly responds, rather than the remote causes of succession involved in a physiographic account of water variations. While Clements did not develop Drude's articulation of the phytogenic relation of ecological formations, he nevertheless appropriated from him the notion of a plant formation as a vegetative unit that is organized by the biological processes that are directly influenced by proximate climate variations.

Cowles and Clements both provided accounts of plant succession that emphasized the dynamic or developmental changes of plant associations in their attempts to answer the question concerning the geographical distribution of plants. They both presented this new explanatory shift in contrast with the static explanations of plant distributions that characterized the 19th century German tradition of plant geography inaugurated by Alexander von Humboldt. The very thematization of succession as a topic of research involved a shift in explanatory interest from static botanical classifications to a genetic interest in the habitat conditions of plant distribution. In other words, they addressed questions concerning the geographical distribution of plants through a focus on variations among plant associations over time rather than a botanical classification of plant associations that focused on the structural characteristics proper to species that persist through temporal variations.

These accounts of plant succession, however, were significantly different with regard to their fundamental assumptions and explanatory principles. We have seen that the logics of habitat associations that were presupposed in these genetic accounts of plant associations were different with regard to the characteristics of the collective attributes of plant associations. The logic of habitat associations in Cowles' account of freshwater dune succession presupposed a basic fit between topological stability and the physiological attributes of individual specific populations. By contrast, the logic of habitat associations in Clements' account of Nebraska prairie succession presupposed a basic fit between climatic conditions (e.g., temperature and precipitation) and functionally self-organized climax communities as a collective whole. The identification of the logical sense of this difference in the collective attributes of plant associations in the work of these two pioneers in American plant ecology clarifies the basic assumptions of their respective classifications of plant associations.

The philosophical differences between Cowles and Clements can also be clarified through a contrast between their respective explanatory principles. The difference between the static classifications of plant associations by 19th century plant geographers and the genetic investigations of succession in Cowles and Clements can also be characterized in terms of an explanatory interest in efficient causation and historical description. In other words, the methodological departure of genetic or dynamic methodologies from the static classifications of the 19th century involved an interest in causal explanations of the natural history of plant associations. The genetic or dynamic explanations of succession not only can be characterized in terms of an interest in temporal variations of dune or prairie habitats, but the efficient causes of these habitat changes. Both Cowles and Clements made an essential scale distinction between the local edaphic and regionally climatic in their respective accounts of succession and they both maintained that there are causal conditions that operate at both levels of analysis. However, they disagreed about the basic causes of succession, namely, about whether these causes primarily operate at the local edaphic scale of plant associations or the regionally climatic scale.

Cowles and Clements both maintained that there are at least two kinds of causal explanations for the phenomena of plant succession – proximate and remote. Clements maintained that the proximate causes of prairie succession were the climate variations of regional temperature and precipitation, on the one hand, and the remote causes of prairie successions were the physiographical variations involved in surface geology and topology, on the other. As he stated,

> In dealing with the causes of development and especially with initial causes, it must be borne in mind that forces of nature are almost inextricably interwoven. In all cases the best scientific method in analysis seems to be to deal with the immediate cause first, and then to trace its origin just as far as it is possible or profitable. Throughout a climax formation, physiography usually produces a large or the larger number of developmental areas. The influence of physiography in this respect is controlled or limited by the climate.... Apart from the gain in clearness of analysis, greater emphasis upon the proximate cause seems warranted by the fact that it is the chresard to which the plant responds, and not the soil-texture or the physiography. In like manner, the invasion of a new area is a direct consequence of the action of the causative process and not of the remote forces behind it. The failure to consider the sequence of causes has produced confusion in the past...and will make more confusion in the future as the complex relations of vegetation and habitat come to be studied intensively.[59]

The proximate causes of variations of plant associations are regional temperature and precipitation variations and the remote causes of plant associations are the physiographic features of a habitat (surface geology and water drainage topology).

The development or evolution of climax formations is most immediately (proximately) conditioned by regional climate variations. In contrast, the causal explanations that appeal to physiographic features of a habitat are mediated through remote geological history. In short, succession is proximately caused by regional climatic variations and remotely caused by the geological and hydrological history (physiography) that conditions differences at the local edaphic scale.

Cowles also made a distinction between proximate and ultimate (remote) explanations of the changes involved in plant succession, but he inversely attributed these kinds of causal explanations of plant succession to the local and regional scale. Cowles stated, "The author [Cowles] feels, in view of the increasing objection to the use of the terms 'edaphic' and 'climatic,' that is may be well to use such terms as 'proximate' and 'ultimate.'"[60] In contrast to Clements, Cowles maintained that the proximate causes of distributive variations of plant associations were the variations of the moisture content of the soil that are primarily and immediately conditioned by the topological variations of water drainage. The proximate causes of succession that condition the variations among "edaphic formations" concern soil moisture variations. In contrast, ultimate or remote explanations of succession involve regional climatic conditions. Explanations of regional "climax formations" in Cowles' view were ultimate or remote in the sense that temperature and precipitation variations are determined by planetary processes related to the earth's rotation around the sun. Causal explanations that seek the ultimate origin of plant succession point to the annual variations that are determined, in turn, by variations in the earth's orbital rotation. These astronomical variations remotely condition changes among plant associations – there is a linked series of mediate or indirect causal explanations from plant association and distribution to succession, climatic conditions, and planetary orbit. By contrast, the explanations of local edaphic formations of plant associations were proximate in the sense that soil moisture content directly or immediately conditions plant distribution. Recall that Cowles' physiography included classification features from both surface geology and topology. While it may be the case that the surface geological factors that condition the distribution and succession of plant associations provide remote or ultimate explanations, the topological factors of habitats in Cowles account provide proximate explanations. As we have seen, Cowles' account of succession did not merely correlate distributive changes in plant associations with various habitat changes. Rather, his classification of plant associations was based on soil moisture tolerance (xerophytic, mesophytic, and hydrophytic) and thematized water variations in the soil as the identity in both the manifolds of topological variations involved in water drainage, e.g., erosion and deposition, and the physiological processes involved in plant nutrition. Causal explanations of the distribution of plant associations at the local edaphic scale are proximate in the sense that water variations directly condition the changes in both topological and plant succession. In short, water variations in the soil are proximate causes of the local distribution, for Cowles, of plant associations through the processes of succession.

1.6 Lake succession

Cedar Creek Bog is currently located in the 2,200-hectare preserve operated by the University of Minnesota and is located approximately 50 km north of Minneapolis and St. Paul. Cedar Creek Preserve has post-glacial soils that can be generally characterized as pitted outwash sand plain – relatively flat sandy soil that is pitted with kettles (topographical depressions). The soil composition spans five of the ten soil orders. Generally speaking, the topology of the Cedar Creek Bog is an intricate pattern of upland and lowland patches and indicates that the central part of the basin east of Second Island and Crone Island is due to the melting of a large block of glacial ice. As this last block of ice melted, it formed a drainage lake more than 12 meters in depth that has been gradually receding to less than a tenth its original depth and area – Cedar Bog Lake. This lake has been considered a classic case of late-stage lake succession since Lindeman's investigations during 1936–1940.

The Anoka Sand Plain of east central Minnesota is comprised of an archipelago of lakes, bogs, and marshes in various stages of hydrarch (shallow water) succession. The succession of these bodies of water into mature forests is conditioned and constrained by late Wisconsin glacial history. This geological history of the Anoka Sand Plain is related to the southwestern migration of the upper Mississippi River. During the late Wisconsin glaciation period, ice from the Keewatin center moved southeastward across central Minnesota and terminated in an encounter with the high terminal moraines of the Patrician drift.[61] The predominant mass of ice pushed southward to form the Des Moines Lobe, but a thin sheet of ice known as the Grantsburg sublobe deflected northeastward across the specific Anoka region. The Grantsburg sublobe advanced northeastward and pushed along the precursor of the present upper Mississippi River watershed. The ice impounded the river to form Lake Grantsburg which eventually found a channel outlet through, what is now, the modern St. Croix River. As the Grantsburg sublobe stagnated and began to recede, the river gradually migrated back across the exposed till and melting ice and continued to empty into the St. Croix channel near Taylor's Falls. As the ice eventually receded to the original upper Mississippi channel, the river reverted to its previous course – the course it has today. This southwestern migration of the river, according to William Cooper, left in its wake large expanses of outwash sand and small locales of level till that included fragments of moraine.

The vegetation of Cedar Creek Preserve is remarkably diverse and includes characteristic species from each of the three large ecosystems of North America (western prairies, northern evergreen forests, and eastern temperate forests). It has a continental climate with significant seasonal variation from cold winters (mean January temperature is −10 degrees Celsius) to warm summers (mean July temperature is 22.2 degrees Celsius). These seasonal temperature fluctuations constrain and condition the regional distribution of these prairies and forests. The mosaic uplands are dominated by oak savanna, hardwood, and pine forests. The lowlands are comprised of ash and cedar swamps, acidic bogs, marshes, and sedge meadows. The regional lowlands, in particular, have filled in much of the pitted basin of kettle-holes and depressions with peat that has been formed largely from sedges. The less well-drained basins have characteristically developed tamarack or spruce bogs. Cedar Bog Lake is one of these less well-drained kettle lakes, but is regionally distinctive in that its bog forest is dominated by white cedar or arbor vitae (*Thuja occidentalis*) that has been gradually encroaching upon the basin, rather than the characteristic tamarack-spruce association.

If primary dune succession illustrates the beginning stages of the terrestrial distribution of plants, the late stages of senescent lakes such as Cedar Bog Lake illustrate the culmination of hydrarch succession. Cedar Bog Lake is a dying lake, if you like. The various serial formations of outer hydrarch succession can initially be indicated by peripheral girdles of vegetation along the margins of the open water. This outer margin is characterized by a bog forest that is peripheral to the sedge mat encircling the open water and occupies the sloping topography of the post-glacial basin of the lake. A tamarack population (*Larix laricina*) about 40 meters in width shapes a distinct girdle that progresses (older to younger individuals) inward and has ground cover that is comprised of various mosses and heaths. This progression is noteworthy not only in that it indicates a definite spatial directionality, but a successional relationship with the white cedars (*Thuja occidentalis*). Tamaracks cannot reproduce in their own shade – they have independent reproduction strategies with regard to spatial distribution. Tamarack seedlings require more shade than their parent trees provide and are abundant in association with mature white cedars that provide more shade. White cedar seedlings, in turn, are more abundant with less shade and characteristically distributed below the mature tamaracks. The individuals of this tamarack-white cedar association in Cedar Bog Lake increase in size and abundance toward the outer edges of the margin.

The inner sedge mat is composed of amphibious plants that are minimally dependent on the support of a substratum. The two dominant species of this encroaching border are swamp loosestrife (*Decodon verticillatus*) and cattail (*Typha latifolia*) which compete for mat establishment. This competitive association cyclically varies according to annual climate factors – *Decodon* is dominant during the years with higher precipitation and lake levels, while *Typha* is dominant during dryer years. The *Decodon-Typha* association is noteworthy in that it accelerates the encroachment of the mat upon the open water of the lake. Lindeman measured a radial distance decrease of approximately one meter every five years and estimated that Cedar Bog Lake would become a bog swale within 250 years.[62]

The lake vegetation is comprised of a variety of pondweeds such as *Najas flexilis*, *Ceratophyllum demersum*, and *Potamogeton zosteriformis* and *Gloeotrichia* that perennially become abundant. Algae blooms occur during certain years and have an incomplete deposition that contributes to the sedimentary ooze and the lower water column. The lake vegetation is conditioned by bottom sandy soil and the accumulated layers of sediment. Lindeman surveyed the lake sediment by drilling 25 cores with a modified Davis borer and graphically represented the

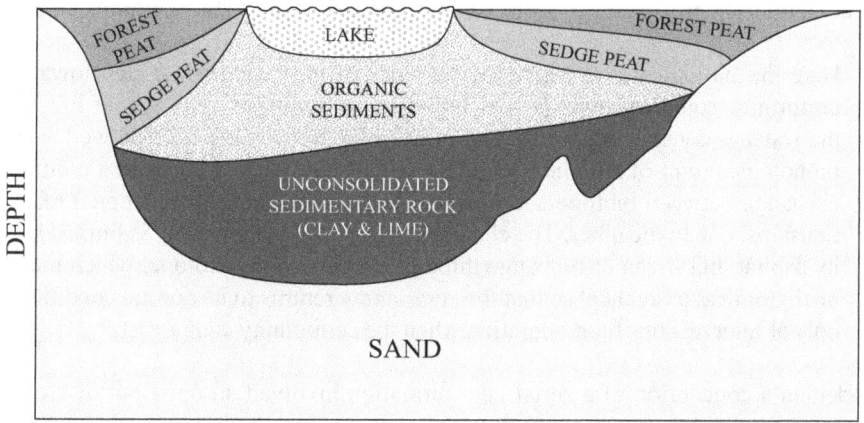

Figure 1.2 Transect profile of Cedar Bog Lake.

Raymond L. Lindeman, "The Developmental History of Cedar Creek Bog, Minnesota," *American Midland Naturalist* 25, no. 1 (1941): 107.

sediment profiles of these borings. These sediment profile samples were taken at depth intervals of every 20 centimeters of the shallower sediment and every 50 centimeters of the deeper layers of sediment (see Figure 1.2). Generally speaking, the sediment profiles were comprised of distinct layers that correspond to the various stages of lake succession, e.g., oligotrophic, mesotrophic, eutrophic, and senescent sediments. The topology of sediments that filled the kettle bottom of this post-glacial lake has an increasing contour from the deepest northwest bottom to the relatively shallow southeastern shore. This bottom topography gains significance in light of the prevailing northwesterly winds that contribute to the high rate of deposition to the southeast. The more extensive sedge and forest peat sediments on the southeastern shore indicate that the succession toward bog-forest associations has had a higher rate of variation. In short, Lindeman's sediment profiles illustrated the physiographic features of the lake habitat and demonstrated successive stages of oligotrophic, mesotrophic, eutrophic, and senescent productivity of aquatic plant associations.

1.7 The logic of habitat associations in Lindeman's account of lake succession

One of the breakthroughs in Lindeman's account of lake succession involved a shift in an understanding of succession in terms of the variation of species composition to the variations involved in food cycles. Lindeman's account of the serial formations involved in hydrarch succession in terms of food chains was a methodological extension of Charles Elton's notion of a food pyramid developed in the context of terrestrial animal interactions and August Thienemann's extension of the notion of "productivity" from community economics. Thienemann stated,

> Here the biological and physiological sides of the conographic step toward uniformity come together in real limnological synthesis. The whole life of the water is considered; the development of lie in the water is conceived as a limnological unit of a higher order. These units come about through a mutual exchange between biotope and biocoenosis, both standing in functional relationships, one to the other. The character of the living community is limited by its habitat, but it can also change through stability of the biotope, which may be rhythmical or cyclical so that the community returns to its normal condition only at intervals or of long duration when the community can persist.[63]

Lindeman's conception of a serial lake formation involved an open but discrete unit of analysis that was scaled in a way that was conducive to the quantitative analysis of the annual transfer of calories through the trophic system. The philosophical significance of this innovative conception of formation lies in the introduction of new methodological principles and fundamental assumptions to address the problems of plant succession. One of these problems concerns the proper characterization of the functional attributes of ecological communities. Lindeman's conception of serial lake formations addressed the problem concerning the functional attributes of community associations through a reduction of phenotypic characteristics of formations to the chemical constituents of nutritive cycles. The new object of investigation became the annual trophic cycle of the lake and the basic unit of analysis became the calorie. This was a decisive methodological breakthrough in 20th century accounts of succession whose logic of habitat associations is still employed in limnology today.

The seasonal stages of the distribution of the calories in a food cycle can be grouped into more or less discrete trophic levels. The initial tropic level in any serial formation involves a transfer of solar energy into nutrients through a process of photosynthesis. Photosynthesis is a process of synthesizing complex organic compounds from simple inorganic substances. In Lindeman's Cedar Bog Lake, the primary trophic producers or "autotrophs" were macrophytic pondweeds and microphytic phytoplankton. Although these plants release a portion of potential energy through respiration and other catabolic processes, a surplus of organic compounds is accumulated and transferred more or less efficiently to primary consumers or "heterotrophs" of Cedar Bog Lake such as tadpoles, ducks, fishes, insects, and various zooplanktons. Similar to terrestrial herbivores, these primary consumers or browsers feed upon this surplus of potential energy. A considerable portion of this potential energy is used for the kinetic energy involved in metabolism of individual organisms, but the remainder of nutrient transfer is stored by the primary consumer. These primary consumers are eventually eaten themselves or die and the energy transfer involved in nutrient distribution is passed on to secondary consumers such as fish crustaceans, turtles, frogs, and birds (see Figure 1.3).

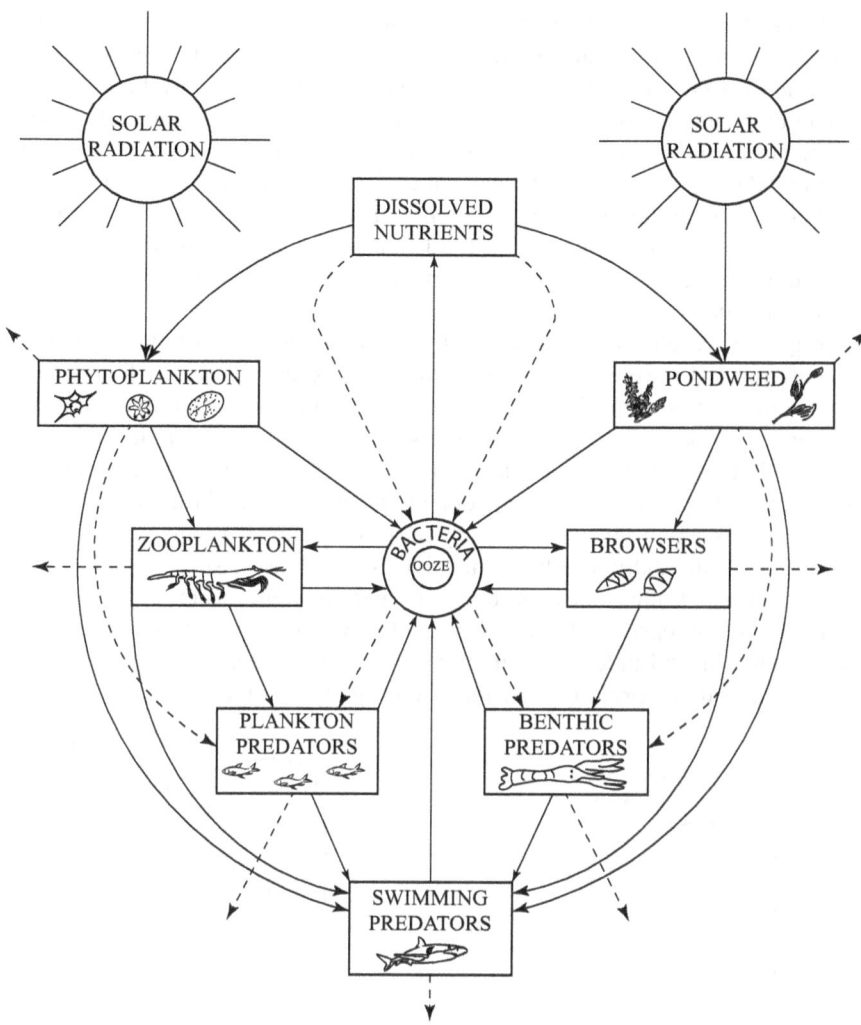

Figure 1.3 Generalized lacustrine food-cycle relationships.

Raymond L. Lindeman, "The Trophic Dynamic Aspect of Ecology," *Ecology* 23, no. 4 (1942): 401.

Lindeman's analysis of the trophic dynamics of Cedar Bog Lake quantified the annual variations in trophic production while correcting for the energy loss due to respiratory and metabolic processes involved in trophic transfers. The shift in explanatory focus to nutrient distribution simplifies the complexity involved in community interaction, e.g., the predator–prey relationships that organized Elton's account of carnivorous trophic levels, and unifies the manifold of chemical variation according to rates of caloric productivity and efficiency of trophic transfer. This shift of theoretical interest changes the object of investigation from the observed abundance of a given species to the production of organic nutrients proper to accumulated biomass at a given trophic level. This shift in theoretical interest to the productivity of trophic levels involves an abstraction from the phenotypic and behavioral attributes of individuals and populations and a reduction of community interactions to their chemical constituents. This methodological advance has a theoretical interest not only in measuring the gross production of the sum total of nutrients stored or spent at each trophic level, but the net production after the energy utilized in respiration is factored. Both productivity figures could be given in gram calories and Lindeman expressed productivity in terms of the number of calories used on a given square meter or centimeter during an annual timescale.

The general results of Lindeman's investigations into the trophic dynamics of Cedar Bog Lake were unexpected. Lindeman expected that the four annual cycles would demonstrate the annual equilibrium of a hydrarch climax association. Chancey Juday's investigations of Lake Mendota, Wisconsin not only provided baselines involved in calculating lake net productivity, but an operative assumption that hydrarch lakes transitioned into bog forests through an uninterrupted eutrophy (high nutrient, low oxygen) of its annual production. Instead, Lindeman's results did not indicate positive correlations of such a transition of serial formations. Rather, these productivity measurements seemed to "vary tremendously and independently" and it was clearly evident that Cedar Bog Lake was not eutrophic.[64] Lindeman characterized its successional stage as senescent (old age) stage after a prolonged period of eutrophy (see Figure 1.4).

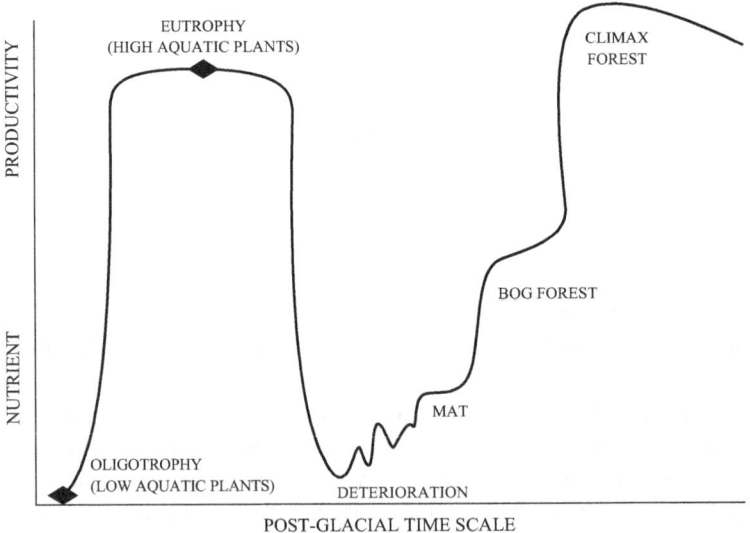

Figure 1.4 Hypothetical productivity growth-curve of a hydrosere, developing from a deep lake to climax in a fertile, cold temperate region.

Raymond L. Lindeman, "The Trophic Dynamic Aspect of Ecology," *Ecology* 23, no. 4 (1942): 413.

The specific results of Lindeman's investigations illustrate innovations with regard to the fundamental assumptions and explanatory principles of inland lake succession. More specifically, the senescent formations of hydrarch succession were conceived through the annual production and efficiency ratios at work in the trophic transfers. Lindeman's specific results supported the general claim that productivity rates and efficiency ratios increase in early stages of succession and consumer efficiencies increase during succession. Consumer efficiencies here refers to how many calories a consumer passes on to other organisms after correcting for respirations and metabolism. Following G. Evelyn Hutchinson, Lindeman set the baseline rate of total possible solar energy entering an ecosystem at 118,872 calories per square centimeter per year and discovered that the producers (pondweeds, phytoplankton, etc.) only captured one-tenth of one percent (0.10%) of solar radiation throughout the year. Cedar Bog Lake's efficiency ratio was significantly lower than the efficiency ratios of producers in Juday's data from Lake Mendota, Wisconsin. After correcting for respiration, metabolism, and decomposition, Lindeman calculated the rate of productivity at 111.3 caloric grams per annual square centimeter that was available for herbivorous consumers in the lake. The transfer of energy to primary consumers was measured from this net productivity of plant producers. The primary consumers (herbivores) had a significantly lower rate of net productivity at 14.8 caloric grams which indicated that the primary consumers were eating less than one seventh of the pondweeds, phytoplankton, and other producers – the remainder was unused or decomposed. Lindeman's value of 3.1 for secondary consumers was the net productivity limited to the small cyprinoid fish in the lake. While these secondary consumers consumed more of the available food than primary consumers, they also used more calories in respiration and the metabolism necessary for predation. While the fish utilized more calories in their metabolic processes, they nevertheless had a higher efficiency ratio relative to primary consumers at 22.3%. These productivity rates and efficiency ratios (along with comparisons from Juday's data from Lake Mendota) provided Lindeman with efficiency quotients that evidenced two main patterns of trophic dynamics. Lindeman called these "fundamental trophic principles."[65] The first principle was the generalization of increasing efficiency – "consumers at progressively higher levels in the foot cycle are progressively more efficient in the use of their food supply."[66] The Cedar Bog efficiency ratios progressively increased from 0.10% for producers to 13.3% for primary consumers and 22.3% for secondary consumers. The second principle of the calculations of caloric energy transfer among trophic levels concerned the loss of energy due to respiration. Caloric energy loss due to respiration in Lindeman's calculations increased from 21% for aquatic plants to 60% for carnivorous secondary and tertiary consumers.

1.8 Conclusion

Each of these accounts of succession in late 19th and early 20th century plant ecology operated with a logic of habitat associations with fundamental assumptions and explanatory principles. The differences among the accounts of succession can be clarified through the part-whole logic of the ecological notion of form. The physiological account of primary dune succession advanced by Warming and Cowles operated with a conception of explanatory necessity that is specific to geographical water variations and organized its classification model accordingly. This conception of necessity was an application of the developments in 19th century plant physiology that explained plant growth in terms of the chemical variations of organic parts, e.g., roots, shoot, and leaves. This breakthrough in 19th century plant physiology played a decisive role in the physiological ecology of Warming and Cowles and could be considered to represent an initial stage of chemical explanations of plant associations. This breakthrough led to the discovery that there are several chemical elements that are considered essential for plant life – nitrogen, phosphorous, potassium, calcium, magnesium, sulfur, boron, chlorine, iron, manganese, zinc, copper, molybdenum, and nickel. Nitrogen is a particularly important component of chlorophyll, the compound through which plants assimilate energy from sunlight to produce sugars from water and carbon dioxide. Phosphorus is also a particularly important element in that it helps plants convert other nutrients into usable building blocks with which to grow. In both cases, plants absorb these nutrients as ions in water. These chemical variations in the nutritive cycle of plants provided 19th century plant physiologists with a way to explain the functional activity involved in nutrition with a strong notion of necessity. Plant nutrition is accomplished through the interdependence of plant assemblages, e.g., roots, shoot, and leaves, and this accomplishment can be physiologically explained through chemical processes of its basal conditions, e.g., cell growth and division.

Warming's physiological approach to the questions concerning the geographical distribution of plants was an application of advances of 19th century plant physiology. As we have seen, Warming and Cowles' accounts of the primary succession of sand dunes utilized the application of the methodological principles and fundamental assumptions of plant physiology to address the variations involved in local or edaphic plant distribution. Lindeman's account of lake succession also applied this physiological conception of necessity to the changes in plant and animal distributions of inland aquatic habitats. Through an analysis of the trophic levels involved in lake succession (Lindeman) and the nitrogen and phosphorus cycles of seasonal lake variations (Hutchinson), this chemical notion of necessity was applied to post-glacier lakes as units of analysis. In this regard, the logic of habitat fitness at work in Lindeman's account of Cedar Bog Lake (Minnesota) can be properly characterized as a physiological account in the sense that the primary unit of analysis was edaphic plant associations geographically determined by the permeable boundaries of a lake.

Lindeman's account of lake succession also addressed the functional conception of necessity that was operative in Drude and Clements's logic of habitat associations. Consider the functional explanation of forest succession from this holistic approach. Forest climax associations such as beech-maple or oak-savanna illustrate self-regulatory processes such as canopy structure and littoral sedimentation in soil compositions that are more or less organized by climate variations such as temperature and precipitation. The regional distribution of plants cannot be fully accounted for without a consideration of the self-regulatory processes proper to forest formations. As we have seen, these forest formations are analogous to the self-regulatory processes of individual organisms. Clements described the directionality of plant formations of serial stages as one-sided and purposive in light of the stable equilibrium of climax associations. This stochastic tendency of forest succession toward a climax equilibrium has a functional notion of necessity that Clements applied to the plant formation as an idealized whole. That is, Clements' account of succession attributed functional properties to the associations of interspecific individuals and populations. By contrast, Lindeman's account of lake succession backfills, so to speak, the principle of purposiveness in Clements idealized holism with the mechanistic explanations of the chemical variations in nitrogen and phosphorus cycles of inland lakes. As we will see later, the "self-regulating mechanisms" involved in the circular causal systems of lake succession advanced a mathematical measurement of habitat fitness. Generally speaking, these accounts of lake succession address the problem of habitat fitness through statistical techniques made available by probability theory. More specifically, the annual fluctuations involved in the phosphorus and nitrogen cycle and the trophic structure of post-glacial freshwater lakes can be quantitatively measured by the differential equations of set theory in a way that avoids or bypasses the problems of functional properties of communities that plagued Clements' account. This represents a decisive breakthrough in the mathematization and chemicalization of nature in 20th century ecology.

Notes

1 Oscar Drude, "The Position of Ecology in Modern Science," in *Congress of Arts and Science: Universal Exposition, St. Louis, 1904*, ed. Howard J. Rogers, trans. Jane Patten (Boston: Houghton, Mifflin and Company, 1906): 179.
2 Henry Chandler Cowles, "The Ecological Relations of the Vegetation on the Sand Dunes of Lake Michigan. Part I – Geographical Relations of the Dune Flores," *Botanical Gazette* 27, no. 2 (1899): 95–117; "The Ecological Relations of the Vegetation on the Sand Dunes of Lake Michigan (Continued)," *Botanical Gazette* 27, no. 2, 3, and 4 (March, April, May 1899): 167–202, 281–308, 361–391; "The Physiographic Ecology of Chicago and Vicinity: A Study of the Origin, Development, and Classification of Plant Societies," *Botanical Gazette* 31, no. 2 (1901): 73–108; "The Physiographic Ecology of Chicago and Vicinity: A Study of the Origin, Development, and Classification of Plant Societies (Concluded)," *Botanical Gazette* 31, no. 3 (1901): 145–182; "The Fundamental Causes of Succession Among Plant Associations," *Report of the British Association for the Advancement of Science, 79th Meeting, Winnipeg: 1909* (London: John Murray, 1910): 668–670.
3 Roscoe Pound and Frederick E. Clements, *The Phytogeography of Nebraska: General Survey*, 2nd ed. (Lincoln: Botanical Seminar, 1900); Clements, Research

Methods in Ecology (Lincoln: University Publishing Company, 1905); Clements, *Plant Succession: An Analysis of the Development of Vegetation* (Washington: Carnegie Institution, 1916).

4 Raymond L. Lindeman, "The Developmental History of Cedar Creek Bog, Minnesota," *American Midland Naturalist* 25, no. 1 (1941): 101–112; Murray F. Buell and Helen Foot Buell, "Surface Level Fluctuations in Cedar Creek Bog," *Ecology* 22 (1941): 314–321; Raymond L. Lindeman, "Seasonal Food-Cycle Dynamics in a Senescent Lake," *American Midland Naturalist* 26, no. 3 (1941): 636–673; "Experimental Simulation of Winter Anaerobiosis," *Ecology* 23, no. 1 (1942): 1–13; "The Trophic Dynamic Aspect of Ecology," *Ecology* 23, no. 4 (1942): 399–417.

5 Frank Benjamin Golley, *A History of the Ecosystem Concept in Ecology: More Than the Sum of the Parts* (New Haven: Yale University Press, 1993), 46.

6 Eugene P. Odum and Gary W. Barrett, *Fundamentals of Ecology*, 5th ed. (Belmont: Thomson, 2005), 4–7.

7 Eugene Warming, *Oecology of Plants: An Introduction to the Study of Plant-Communities* (Oxford: Clarendon, 1909): 263.

8 Cowles, "Ecological Relations," 182.

9 Ibid.

10 Cowles, "Ecological Relations," 184.

11 Ibid., 184.

12 Ibid., 186.

13 For further discussion of the historical trajectory of Cowles' physiographic ecology, see Gregory J. Cooper, *The Science of the Struggle for Existence: On the Foundations of Ecology* (Cambridge: Cambridge University Press, 2003), 37–42; Golley, *History of the Ecosystem Concept in Ecology*, 11–15; Robert P. McIntosh, *The Background of Ecology: Concept and Theory* (Cambridge: Cambridge University Press, 1985), 43–49; Ronald Tobey, *Saving the Prairies: The Life Cycle of the Founding School of American Plant Ecology, 1895–1955* (Berkeley: University of California Press, 1981), 106–109; Donald Wooster. *Nature's Economy: A History of Ecological Ideas*, 2nd ed. (Cambridge: Cambridge University Press, 1977), 206–208.

14 Warming, *Oecology of Plants*, 40. Warming further stressed the importance of soil moisture content, "The oecological importance of water to the plant is fundamental and almost surpasses that of light and heat," Ibid., 28.

15 Ibid., 44.

16 Cowles, "Ecological Relations," 75.

17 Warming, *Oecology of Plants*, 14.

18 Cowles, "Ecological Relations," 78.

19 Rogers and Roberton highlight this physiographic aspect of Cowles' approach, "Throughout his analysis of succession on the dunes, Cowles described the positive interaction of plants and dunes in which each may alternately control the character of the other. His chief contribution in these works was his elucidation of the various ways in which physiographic processes control biological ones." See Garry F. Rogers and John M. Robertson "Henry Chandler Cowles: 1869-1939," *Geographers Biobibliographical Studies* 10 (1986): 30.

20 Cowles, "Physiographic Ecology of Chicago," 79.

21 Cowles physiographic classification of plant societies in general proceeded by a general distinction between inland and coastal groups. The inland group was comprised of a (1) river series (ravine, river bluff, flood plain), (2) pond-swamp-prairie series (pond, undrained swamp, prairie), and (3) upland series (rock hill, clay hill, sand hill). The coastal group was comprised of (1) lake bluff series, and (2) beach, dune, sand hill series (beach, embryonic beach, active or wandering dunes). Cowles develops this physiographical classification in "Physiographic Ecology of Chicago," 86–108, 145–177.

22 Ibid., 112.
23 Cowles, "Ecological Relations," 111.
24 Ibid.
25 Warming, *Oecology of Plants*, 12.
26 Ibid., 92.
27 Ibid., 95.
28 Ibid., 81.
29 Cowles, "Fundamental Causes of Succession," 668.
30 Cowles, "Physiographic Ecology of Chicago," 81.
31 Ibid., 83. Cowles continued, "In a study of plant societies such as this, it must be recognized that orders of succession are not the same in various regions," Ibid., 82.
32 Cowles, "Ecological Relations," 96.
33 Henry Chandler Cowles, "Present Problems in Plant Ecology: The Trend of Ecological Philosophy," *American Naturalist* 43 (1909): 357.
34 John Bradley, "Travels in the Interior of North America, ca. 1815," in *Early Western Travels: 1748–1846*, Vol. 5, ed. Reuben Gold Thwaites, trans. H. E. Lloyd (Cleveland: Arthur H. Clark Co., 1904).
35 Frederick Clements, "Nature and Structure of the Climax," *The Journal of Ecology* 24 (1936): 255.
36 Ibid., 273.
37 Ibid., 255.
38 Ibid., 265.
39 Cowles agreed with this characterization of prairie associations as the climax association of the North American Great Basin, "east of Manitoba all other associations tend to develop into mesophytic conifer forest whereas to the westward all other associations tend to develop into a prairie." See Cowles, "Fundamental Causes of Succession," 669.
40 Clements, *Plant Succession*, 6.
41 Frederick Clements, *Research Methods in Ecology* (Lincoln: University Publishing Co., 1905): 173–175; Tobey, *Saving the Prairies*, 57–75. Technically speaking, the experimental design of denuded quadrats measures the secondary stages of succession in that its thematic object is the re-population of the quadrat from the surrounding prairie associations.
42 Clements, *Plant Succession*, 4. For historical treatments of Clements' account of succession, see Tobey, *Saving the Prairies*, 76–98; McIntosh, *The Background of Ecology*; Cooper, *The Science of the Struggle for Existence*, 37–42; Wooster, *Nature's Economy*, 208–220.
43 Ibid., 3–7. Clements, "Nature and Structure of the Climax," 252–284.
44 Clements, *Research Methods*, 199; see also *Plant Succession*, 3.
45 Ibid.
46 Ibid., 6.
47 Ibid., 145. See also Clements, *Phytogeography of Nebraska*, 49.
48 Ibid., 5.
49 Ibid., 6.
50 Clements, *Phytogeography of Nebraska*, 14.
51 Drude, "Position of Ecology," 180. Drude also methodologically characterized this phrase as "the methods of the doctrine of formations" (183).
52 Ibid., 179.
53 Ibid., 184.
54 Clements' account of succession operated with a notion of a regional plant formation as a complex super-organism and his notion of plant formation was indebted to August Grisebach's notion of vegetative form. In the following chapter, we will see how Warming distinguished his notion of growth form from the "vegetative form" introduced by Grisebach and employed in systematic botany to classify the diversity

of floral structures and methods of pollination. Grisebach introduced the term "formation" in 1838 in an effort to incorporate recent developments in historical geography into Humboldt's account of regional physiognomy and, in the process, demonstrated how Humboldt's account of regional physiognomy correlates climatic influence with the morphological form of various plant species. See Grisebach's *"Über den Einfluss des Klimas auf die Begrenzung der natürlichen Floren,"* reprinted in *Gesammelte Abhandlungen und Kleinere Schriften zur Pflanzengeographie* (Leipzig: Verlag von Wilhelm Engelmann, 1880): 1–29; *"Pflanzengeographie und Botanik,"* in *Alexander von Humboldt*, vol. 3, edited by Karl C. Bruhns (Leipzig: Brockhaus, 1872): 248.

55 Drude, "Position of Ecology," 184.
56 Ibid., 186.
57 Ibid., 186.
58 Ibid., 183.
59 Clements, *Plant Succession*, 6.
60 Cowles, "Fundamental Causes of Succession," 669.
61 William S. Cooper, "The History of the Upper Mississippi River in Late Wisconsin and Post-glacial Time," *Minnesota Geological Survey Bulletin* 26 (1935): 1–116.
62 Lindeman, "Developmental History of Cedar Creek Bog," 105.
63 August Thienemann, *Die Binnengewässer Mitteleuropas* (Stuttgart: E. Schweizerbart'sche Verlagsbuchhandlung, 1925): 20–22. English translation from Frank B. Golley, *History of Ecosystem Concept*, 39. Lindeman characterized Thienemann's logic of habitat fitness: "This constant organic-inorganic cycle of nutritive substance is so completely integrated that to consider even such a unit as a lake primarily as a biotic community appears to force a 'biological' emphasis upon a more basic functional organization." Lindeman, "Trophic-Dynamic Aspect of Ecology," 400.
64 Lindeman, "Experimental Simulation of Winter Anaerobiosis in a Senescent Lake," 1.
65 Lindeman, "Trophic Dynamic Aspect of Ecology," 407.
66 Ibid., 407.

Bibliography

Bradley, John. "Travels in the Interior of North America, ca. 1815." In *Early Western Travels: 1748–1846*, Vol. 5, edited by Reuben Gold Thwaites, translated by H. E. Lloyd. Cleveland: Arthur H. Clark Co., 1904.

Buell, Murray F. and Buell, Helen Foot. "Surface Level Fluctuations in Cedar Creek Bog." *Ecology* 22 (1941): 314–321.

Clements, Frederick E. *Research Methods in Ecology*. Lincoln: University Publishing Company, 1905.

Clements, Frederick E. *Plant Succession: An Analysis of the Development of Vegetation*. Washington: Carnegie Institution, 1916.

Clements, Frederick E. "Nature and Structure of the Climax." *The Journal of Ecology* 24 (1936): 252–284.

Cooper, Gregory J. *The Science of the Struggle for Existence: On the Foundations of Ecology*. Cambridge: Cambridge University Press, 2003.

Cooper, William S. "The History of the Upper Mississippi River in Late Wisconsin and Post-Glacial Time." *Minnesota Geological Survey Bulletin* 26 (1935): 1–116.

Cowles, Henry Chandler. "The Ecological Relations of the Vegetation on the Sand Dunes of Lake Michigan. Part I – Geographical Relations of the Dune Flores." *Botanical Gazette* 27, no. 2 (1899a): 95–117.

Cowles, Henry Chandler. "The Ecological Relations of the Vegetation on the Sand Dunes of Lake Michigan (Continued)." *Botanical Gazette* 27, no. 2, 3, and 4 (March, April, May 1899b): 167–202.

Cowles, Henry Chandler. "The Ecological Relations of the Vegetation on the Sand Dunes of Lake Michigan (Continued)." *Botanical Gazette* 27, no. 2, 3, and 4 (March, April, May 1899c): 281–308.

Cowles, Henry Chandler. "The Ecological Relations of the Vegetation on the Sand Dunes of Lake Michigan (Continued)." *Botanical Gazette* 27, no. 2, 3, and 4 (March, April, May 1899d): 361–391.

Cowles, Henry Chandler. "The Physiographic Ecology of Chicago and Vicinity: A Study of the Origin, Development, and Classification of Plant Societies." *Botanical Gazette* 31, no. 2 (1901a): 73–108.

Cowles, Henry Chandler. "The Physiographic Ecology of Chicago and Vicinity: A Study of the Origin, Development, and Classification of Plant Societies (Concluded)." *Botanical Gazette* 31, no. 3 (1901b): 145–182.

Cowles, Henry Chandler. "Present Problems in Plant Ecology: The Trend of Ecological Philosophy." *American Naturalist* 43 (1909): 356–368.

Cowles, Henry Chandler. "The Fundamental Causes of Succession among Plant Associations." In *Report of the British Association for the Advancement of Science, 79th Meeting*, Winnipeg, 1909, London: John Murray, 1910, 668–670.

Drude, Oscar. "The Position of Ecology in Modern Science." In *Congress of Arts and Science: Universal Exposition, St. Louis, 1904*, edited by Howard J. Rogers, translated by Jane Patten, 179–190. Boston: Houghton, Mifflin and Company, 1906.

Golley, Frank Benjamin. *A History of the Ecosystem Concept in Ecology: More Than the Sum of the Parts*. New Haven: Yale University Press, 1993.

Grisebach, August. "*Pflanzengeographie und Botanik.*" In *Alexander von Humboldt*, Vol. 3, edited by Karl C. Bruhns, 232–268. Leipzig: Brockhaus, 1872.

Grisebach, August. *Über den Einfluss das Klima auf die Begrenzung der natürlichen Floren*, reprinted in *Gesammelte Abhandlungen und Kleinere Schriften zur Pflanzengeographie*, 1–29. Leipzig: Verlang von Wilhelm Engelmann, 1880.

Lindeman, Raymond L. "The Developmental History of Cedar Creek Bog, Minnesota." *American Midland Naturalist* 25, no. 1 (1941a): 101–112.

Lindeman, Raymond L. "Seasonal Food-Cycle Dynamics in a Senescent Lake." *American Midland Naturalist* 26, no. 3 (1941b): 636–673.

Lindeman, Raymond L. "Experimental Simulation of Winter Anaerobiosis." *Ecology* 23, no. 1 (1942a): 1–13.

Lindeman, Raymond L. "The Trophic Dynamic Aspect of Ecology." *Ecology* 23, no. 4 (1942b): 399–417.

McIntosh, Robert P. *The Background of Ecology: Concept and Theory*. Cambridge: Cambridge University Press, 1985.

Odum, Eugene P. and Barrett, Gary W. *Fundamentals of Ecology*, 5th ed. Belmont: Thomson, 2005.

Pound, Roscoe and Clements, Frederick E. *The Phytogeography of Nebraska: General Survey*, 2nd ed. Lincoln: Botanical Seminar, 1900.

Rogers, Garry F. and Robertson, Jonathan M. "Henry Chandler Cowles: 1869–1939." *Geographers Biobibliographical Studies* 10 (1986): 29–33.

Thienemann, August. *Die Binnengewässer Mitteleuropas*. Stuttgart: E. Schweizerbart'sche Verlagsbuchhandlung, 1925.

Tobey, Ronald. *Saving the Prairies: The Life Cycle of the Founding School of American Plant Ecology, 1895–1955*. Berkeley: University of California Press, 1981.

Warming, Eugene. *Oecology of Plants: An Introduction to the Study of Plant-Communities*. Oxford: Clarendon, 1909.

Wooster, Donald. *Nature's Economy: A History of Ecological Ideas*, 2nd ed. Cambridge: Cambridge University Press, 1977.

2   Logics of habitat fitness

A genealogy of 19th century plant geography

2.1 Introduction

What are the physical laws that organize the geographical distribution of plants? This was the basic question of 19th century plant geography and it was addressed through descriptive classifications that supposed a logic of habitat fitness proper to plant associations. Alexander von Humboldt's "plant forms," Oscar Drude's "vegetative formations," Eugene Warming's "growth form" – each of these notions of plant associations operated with distinguishable logics of habitat fitness that attempted to navigate the methodological "borderland between biology and geography."[1] These logics of habitat fitness were classifications that reflected underlying suppositions concerning plant formations and how they are known. These logics of habitat fitness illustrate a search for a notion of form proper to plant associations in 19th century plant geography. This search for a notion of plant formation can be particularly illustrated through Humboldt's physiognomic classifications of plant associations and organic conception of plant form, on the one hand, and Warming's physiographic classification of plant associations and nutritive conception of plant form, on the other hand. These classifications and conceptions of plant associations proposed fundamental assumptions and explanatory principles that addressed the question concerning the geographical distribution of plants. This chapter is a genealogical investigation of these assumptions and principles that attempts to gain an understanding of the philosophical difference between Humboldt's physiognomic "plant forms" and Warming's physiographic "growth forms."

The distinguishable conceptions of plant formations in Humboldt's physiognomic plant geography and Warming's physiographic plant ecology can be illustrated both historically and logically. Historically speaking, the previous investigation highlighted that different logics of habitat associations were operative in early 20th century approaches to plant succession in American plant ecology. More specifically, Henry Chandler Cowles Cowles' account of primary dune succession was an extension of Warming's notion of growth form and physiological approach to plant geography. Frederick Clements' account of the role of climax communities in plant succession supposed a notion of plant formation that was an extension of physiognomic accounts of plant associations inspired by Humboldt, most notably – Drude's conception of vegetative form. These conceptions of plant formations influenced the fundamental assumptions and methodological principles of the early 20th century debates between the Chicago and Nebraska schools concerning plant succession. I argue that the 19th century static accounts of plant associations were supposed by early 20th century dynamic accounts of plant succession. Cowles made this supposition explicit through a methodological analogy between structural and dynamic geology, on the one hand, and structural botany and ecology, on the other.[2] The genealogy of dynamic conceptions of habitat associations in the accounts of succession in early American plant ecology in the previous investigation raises historical questions concerning the notion of plant form in 19th century European plant geography. I generally agree with Ronald Tobey and Frank B. Golley's historical claim that there were at least two different transatlantic traditions in plant geography.[3] I submit that these distinguishable traditions provided different philosophical conceptions of plant formations that were essentially appropriated in the dynamic methodologies of the pioneering American plant ecologists. This investigation deepens and extends the distinction between two alternative traditions in 19th century plant geography by clarifying the fundamental assumptions and explanatory principles of the respective logics of habitat fitness and conceptions of plant formations.

Second, this investigation identifies and clarifies the logical difference between Humboldt's physiognomic and Warming's ecological plant geographies. Humboldt's pioneering logic of habitat associations supposed a botanical classification of regional plant associations, a Post-Kantian conception of organic form, and a geometric conception of spatial distribution. More specifically, Humboldt's physiognomic logic of habitat fitness prioritized the synthetic achievements of the imagination in accounting for the unity proper to individual plants and plant associations. I argue that Humboldt's part-whole logic and implicit theory of manifolds can be properly characterized as an epistemological idealism that relies on a one-sided emphasis on the synthetic achievements of the plant geographer to account for the unity proper to plant formations. In contrast, Warming's logic of habitat associations illustrates a nutritive conception of plant forms and a different theory of manifolds. Warming's logic of habitat fitness relied on a nutritive conception of plant forms that was an application of discoveries in 19th century plant physiology concerning the role of water in the process of photosynthesis. Warming's pioneering application of a nutritive conception of plant forms to questions concerning the geographical distribution of plant associations can be illustrated through his notion of "growth form" which he develops in direct contrast with the Humboldtian tradition in 19th century plant geography. More specifically, Warming operated with a theory of manifolds that thematized water variations as the basic identity among the manifolds of geographic and nutritive changes. This conception of unity in manifolds of plant geographical change can be properly characterized as an epistemological realism that prioritized the ideographic particularity of given habitats. This investigation reconstructs Warming's logic of habitat fitness and illustrates how this breakthrough was a departure from the fundamental assumptions and explanatory principles of 19th century plant geography.

2.2 Humboldt's physionomic logic of habitats

Humboldt's physiognomic approach to plant associations operated with a logic of habitat fitness that contained ontological suppositions regarding organic parts and wholes. This logic of habitat fitness is perhaps most succinctly articulated in his essay "Ideas for a Physiognomy of Plants." Humboldt's classification of nineteen plant forms is preceded by an extended argument concerning plant associations and the objects of plant geography, the new science Humboldt proposed for the first time. This argument is succinctly stated in the first lines of the essay,

> When a person possessed of an active mind explores Nature, or ponders in imagination the broad range of organic creation, no single one among the manifold impressions that occur to him has so deep and powerful an effect as that of the ubiquitous abundance of life. Everywhere, even near the ice-capped poles, the air rings with the songs of birds or the drone of buzzing insects.[4]

From North to South Pole, from mountain height to ocean depth, organisms cover the earth. Life on planet earth is ubiquitous – it has the generalization proper to the logical category of totality. The science of botany not only has the classification of differences among species as its proper object of investigation, but a concrete unified totality that can be mathematically measured and therefore known with a necessity. Whereas botany was traditionally concerned with the differences among the morphological features of plants, Humboldt gazed upon the new world of the Americas and saw a unity amidst diversity (harmony) that stressed the mutual interdependence at work in the geographical distribution of plants. This unity is known through its physiognomy – a kind of landscape interpretation that is different from, but complementary to, mechanistic explanations. Physiognomy literally means the "knowledge" or "intelligibility" of the "face" or "sensible shape" and this etymology indicates Humboldt's notion of form. He maintained that landscapes (external faces of habitats) have intelligible features that reveal plant associations that can be classified according to a Linnaean binomial taxonomy. The overall unity of a physiognomic account of landscapes is evident in the imagination, e.g., visual impressions and aesthetic experiences, that can be represented in *Naturgemälde* (literally, "nature pictures") or *tableau physique* (landscape images). Analogous to the way in which botanical drawings illustrate the structural features of the various species, landscape painters can illustrate the physiognomy of plant associations (see Figure 2.1).

The physiognomy of regional plant forms can be determined by the "overall impression of a region"[5] that is primarily disclosed through the vegetative cover. Humboldt stated, "the primary determining factor of this impression is the

Figure 2.1 Geography of equinoctial plants. Physical tableau of the Andes and neighboring
     countries according to the observations and measurements done on location
     from the 10th degree of boreal latitude to the 10th degree of austral latitude in
     1799, 1800, 1801, 1802, 1803.

Alexander von Humboldt, *Essay on the Geography of Plants*, ed. Stephen T. Jackson, trans. Sylvie
Romanowski (Chicago: University of Chicago Press, 2009), 145.

covering of vegetation."[6] The vegetative cover of landscapes is such that "one
organic web rests in layers upon layers"[7] such that the overall organic unity is
analogous to the physiognomy of individual organisms. He continued,

> each zone has its own character. The old and profound power of organization,
> despite a certain liberty in the abnormal development of specific cases, binds
> all animal and vegetable life forms to firm, perpetually returning types. In
> the same way that one discerns a certain physiognomy in individual organic
> beings, just as descriptive botany and zoology, in the strict sense of the word,
> are the animal and plant forms, so too is there a physiognomy of Nature that
> applies, without exception, to each section of the Earth.[8]

Descriptive botany and zoology provide accounts of the sensible shape of individual organisms, e.g., leaves, stems, and root structures of plants. Botanical classification explains the differences among plant individuals according to the sensible
shape of organic parts. The botanist can illustrate these sensible shapes through
botanical drawings, that is, they can be visually represented in a way that com-
municates their sensible (and more precisely, visual) shape. The botanical notion
of form as sensible shape provides the explanatory datum that establishes the

logic of difference among plant species. Carl Linnaeus did not change this philo-sophical assumption and methodological principle with his *Systema Natura*, but rather he focused this conception of botanical form to a particular organic part – the reproductive organs. Linnaean taxonomy nevertheless operates with a notion of form as sensible shape; it merely restricts the notion of form to the floristic structures of the plant (specifically – the quantity, size, and shape of reproductive organs). Humboldt's physiognomic logic operated with a notion of form as sen-sible shape that is broader than the Linnaean focus, but he nevertheless appropri-ated a binomial taxonomy that classified types according to primary genera and species. Humboldt emphasized leaf size and shape of the dominant species of the plant association and the tradition of botanical plant geography that followed classified plant associations according to the two primarily dominant species of the plant association. Humboldt's radical innovation was to apply this notion of botanical form (and the classification of difference that it implied) to regional landscapes. This extended quotation indicates his conception of the succession of newly formed volcanic islands:

> as soon as the air first touches the naked stone in the northern countries, webs of silklike fibers form that look like colorful spots to the naked eye. Several of the spots are bordered by exquisite lines, sometimes single and sometimes double; some are cut through with small furrows dividing them into boxes. Their light color darkens with increased age. The vividly bright yellow turns brown, and the bluish-gray of the Lepraria transforms itself gradually to a dusty black. The borders of the aging covering flow into each other and on the dark ground form new, circular lichens of brilliant white. In this way, one organic web rests in layers upon another, and just as the colonizing group of humans must pass through certain stages of moral development, so too is the gradual spread of plants bound by physical laws. Where tall forest trees now lift their tops to lofty heights, there the soiless stone was once covered in such delicate lichens. Mosses, grasses, herbaceous plants, and shrubs fill the gulf of the long but unmeasured era in between. What lichens and moss accom-plish in the North is done in the tropics by *Portulaca, Gomphrenae,* and other oily shore plants. Thus, the history of the covering of plants and their gradual spread over the barren crust of the Earth has its epochs, as does the history of the migrating animal world.[9]

In many senses, Humboldt's physiognomy of plant associations anticipated some of the basic insights of gestalt psychology that subsequently developed in the early 20th century. Like Humboldt, gestalt psychologists such as Kurt Koffka operated with a notion of form as sensible shape through their account of the visual recognition of global percepts or gestalts. Koffka crystalizes the formal fea-tures of gestalts through his well-known and overdetermined phrase "the whole is more than the sum of its parts."[10] Koffka maintained that visual perception oper-ated with an internal relation between figure and ground and he characterized this overall unity as a gestalt. Like Koffka, Humboldt was primarily interested in the overall associative unity proper to the category of totality. In Chapter Three, I will trace this logic out through an engagement with the formal logic of Kant's account of organisms and the explanatory principle proper to reflective judgments. As I develop further through this engagement, Kant's distinction between determinate and reflective judgments is made through an appeal to the logical category of totality. I submit at this point that Kant's logic of organic form is operative in Humboldt's physiognomy of plant associations in at least three senses:

1) plant associations have interdependent parts and wholes that are organized, not merely aggregated,
2) reciprocal relations among mutually dependent parts, and
3) means-end or functional attributes that characterize the reciprocal relations among organic parts and the activity of wholes.[11]

How does Humboldt's account of a physiognomy of plant forms extend this notion of organic forms to collectives of plant species? How are plant associations organized (not merely aggregated) through reciprocal relations with functional characteristics? In what sense does Humboldt's conception of plant form suppose a spatially distributed conception of organic form in the Kantian sense? What is the physiognomic logic of habitat fitness that is operative here?

Humboldt's answers to these questions lay in the interdisciplinary methodology of his general project,

> If one wishes to rise to a general understanding of the views concerning the various forms of life, then my ideas regarding physiognomy, the study of numeric ratios (the arithmetic of botany), and the geography of plants (the study of spatial zones of distribution) cannot be separated from one another.[12]

Taking these three spheres of inquiry into consideration, Humboldt's conception of plant form can be reconstructed according to the following suppositions:

1) There is an overall unity of associative similarity (harmony) to the geographical distribution of plants.
2) Geographical phenomenon are geometrically expressed according to latitude and altitude. The geographical distribution of plants can be quantitatively measured and thereby can be known according to a mathematical necessity.
3) Geographical and vegetative phenomenon are united in a landscape and are distinguished by climate conditions (temperature and atmospheric moisture). The plant forms of vegetative phenomenon are conditioned by climate variations, whereas geography is conditioned by geological history.
4) Isotherms provide the mathematical measurement of regional temperature variations which fundamentally organize the regional distribution of plants.

5) Regional plant forms are primarily constrained and conditioned by climate variations. Similar plant forms are correlated with similar climate conditions (even more so than latitude and altitude determinations).
6) In some cases (especially temperate regions) plant forms are "socially organized" and have holistic properties.
7) The geographical distribution of plants is evident in the landscape physiognomy of plant forms.
8) Individual and collective plants are organisms with organic forms in the Kantian sense.
9) Individual plants can be described through Linnaean taxonomy and plant associations can likewise be described through Humboldt's pioneering classification of 19 plant forms.

These suppositions underlie the logic of habitat fitness in Humboldt's physiognomy of plant forms. This logic is especially noteworthy in its extension of the notion of organic form to plant associations as holistic collectives. Humboldt's conception of organic form was informed by the romanticism of Friedrich Schiller and Johann Wolfgang von Goethe and the organicism of Johann Friedrich Blumenbach, Humboldt's tutor in Göttingen. These romantic and organicist conceptions of organic form were philosophically articulated in the second division of Kant's *Critique of Judgment* (1790). Kant argued in the Third Critique that organic forms are known through reflective judgments. Reflective judgments concerning organic forms can be distinguished from the mechanistic explanations of determinate judgments by an *a priori* principle that is distinctive to the faculty of judgment – the *a priori* principle of purposiveness. For Kant, judgments concerning the reciprocal means-end relations among organic parts and wholes are logically grounded in a principle of purposiveness. This means that organic parts have means-end relations with other parts and the activity of the whole. The distinctive features of living matter concern the reciprocal means-end relations of organic parts and wholes and these organic forms have a necessity that is grounded in the *a priori* principle of purposiveness that is proper to reflective judgments.

I will critically analyze this Kantian conception of organic form in a later chapter, but allow me at this point to simply highlight the sense in which Humboldt's physiognomic logic of habitat fitness operated with a Post-Kantian conception of organic form.[13] The point here is that Humboldt extended the notion of organic form to plant collectives by appealing to the way in which holistic properties of organisms are grounded in the synthetic achievements of the cognizing subject. The unity of organic forms is grounded in and made possible through an *a priori* principle of purposiveness that, according to Kant, cannot be schematized in a determinate judgment. This means that the purposive unity of organic forms – and by analogy plant collectives – obtains its logical necessity in the achievements of the cognizing subject, namely, the visual impressions and aesthetic experience of the plant geographer. The expertise involved in physiognomy concerns the recognition of typical plant associations in the regional landscape and, in particular, the similarities among the botanical types.

The physiognomy of plants should not remain exclusively concerned with the noticeable contrasts in form that the large organisms present when considered individually; it should dare to come to a recognition of the laws that determine the physiognomy of Nature in general, the scenic vegetational character of the entire surface of the Earth, and the vital impression evoked by the aggregation of contrasting forms in various zones of latitude and elevation. Only when examined from this point of view does it become clear wherein lies the close, interior interlinking of the material dealt with in the preceding pages.[14]

The plant geographer is not merely interested in the predominant individuals of the plant association, but the associative unity (similarity) that is given in visual representations. The plant geographer, for example, recognizes beech-maple forest associations, tall grass prairies, and recently glaciated lakes that have characteristics that can be generalized across particular instances. Successful landscape painters can illustrate this associative unity in their best paintings. The point is that Humboldt generalized the notion of organic form to plant associations and conceived of the overall unity of vegetative life on earth as an aggregated patchwork of geographical regions that are geometrically expressed according to latitude and altitude. The geographical distribution of plants is a spatial distribution in such a way that it can be quantitatively measured and known with necessity.

Humboldt's early conception of organic form can be initially indicated through his distinction between animate and inanimate matter in *Florae Fribergensis* (1792). Generally speaking, this work pioneers an account of the geographical distribution of plants prior to Humboldt's travels to the New World (1799–1804) and included an appended series of aphorisms on plant and animal physiology. The distinction between animate and inanimate matter in these aphorisms indicate Humboldt's conception of organic form:

> We call that matter inactive, brute, or inanimate if its elements are combined according to the laws of chemical affinity. We call those bodies animated and organic that, though they tend constantly to change into new forms, are constrained by some internal force, so that they do not relinquish that form originally introduced.... That internal force (*vim internam*) which dissolves the bonds of chemical affinity and prevents the elements of bodies from freely uniting, we call vital.[15]

Organic forms persist through organic variations (e.g., growth) through an internal, vital force which is expressed through excitability or irritability. Vital organs respond to stimuli through sensation and irritable contractions. These excitability and irritability responses indicate a dynamic tension that precludes the bonding of chemical affinities. The internal force is opposed to and prevents the bonding of chemical affinities in the vital organs and prevents chemical bonding from ossification, as it were, and the subsequent loss of reciprocal means-end relations among organic parts and wholes.

The conception of organic form in Humboldt's physiognomic logic can be further illustrated through Humboldt's early (1794–1797) experimental work in Jena on animal electricity. In particular, the first of volume of *Versuche über die gereizte Muskel- und Nervenfaser* (Experiments on Stimulated Muscle and Nerve Fibers) contains his galvanic studies on frogs which tested hypotheses concerning the role of electricity in physiological movements.[16] Humboldt was impressed by the experimental approaches to animal electricity of Bolognese physician Luigi Galvani (1737–1798) who discovered that he could contract leg muscles of decapitated frogs through stimulating the leg muscles with electrical current.[17] More specifically, Galvani's experiment contracted the frog leg muscles by holding the exposed crural nerve near a sparking electrostatic machine and thereby instigated the leg contraction. Galvani conducted a series of experiments to generate a theory of animal electricity in which organisms contained an electrical fluid in muscle fibers. He postulated that the muscle fibers have a positive and negative electrical charge toward a magnetic equilibrium that can be manipulated to produce muscle contraction.

Galvani's characterization of the self-regulation of electric equilibrium in terms of an internal vital force was challenged by Alessandro Volta (1745–1827) who falsified Galvani's theory through a critique of his experimental design.[18] According to Volta, the electrical current that stimulated muscle contractions did not reside from endogenous sources of the frog muscle fibers, but the heterogeneous metals used in Galvani's experiments. It was the metal variations in the hook and metal arcs that produced the current, not the positive and negative charges of the electrical fluid in the muscle fibers. Volta contended that Galvani's experiments did not provide experimental evidence for an internal, vital force (see Figure 2.2).

Humboldt entered into this debate between Galvani and Volta and developed his own experiments on the irritable contraction of muscles. The first volume of Humboldt's research followed Galvani's characterization of the internal force of animals in terms of electro-chemical current and generally attempted to advance Galvani's conception of a vital force. Humboldt revised his account in the second volume (1799) by dropping the reference to an internal force and articulating organic form in Kantian terms.

> I call that stuff animated whose parts, when arbitrarily separated from the whole, change their composition while maintaining their previous external relationships.... The balance of elements in living matter maintains itself only so long as every part is part of a whole and exists as such...in an organism every part is mutually means and end.[19]

In short, Humboldt's investigations of the excitability and irritability of animal physiology illustrates a conception of organic form in the Kantian sense: 1) organisms have interdependent parts and wholes that are organized, not merely aggregated, 2) reciprocal relations among mutually dependent parts, and 3) means-end or purposive attributes that characterize the reciprocal relations among organic

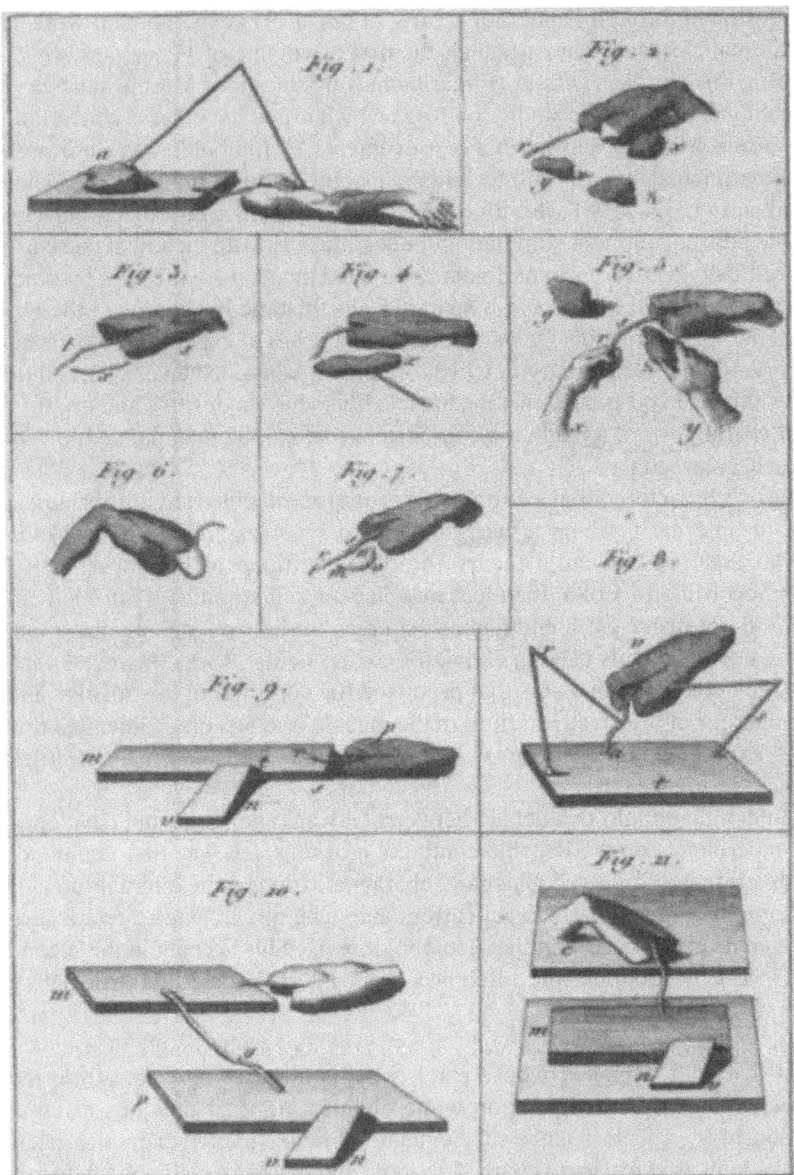

Figure 2.2 Humboldt's illustrations of experiments that demonstrated that excised muscle and nerve tissue arranged in a circuit could produce a current that generates a muscle contraction.

Alexander von Humboldt, *Versuche über die gereizte Muskel- und Nervenfaser nebst Vermutungen über den chemischen Process des Lebens in der Tier- und Pflanzenwelt*, 2. Vol (Berlin: Heinrich August Rottmann, 1797–99).

parts and the activity of wholes. The development of Humboldt's engagement in the debates between Galvani and Volta regarding muscle irritability illustrates that Kant's account of organic form provided a departure from vitalistic explanations of animal electricity. Rather than positing a physical force with magnetic and electrical attributes that applies to animate material things such as frogs, Humboldt's revised approach appealed to a logical account of organic form that did not posit a vital force. The development of Humboldt's early approach to animal electricity provides a unique insight into the basic assumptions and explanatory principles at work in his conception of organic form. In what sense is this conception of organic form applied to the collective attributes of plant associations in Humboldt's pioneering account of plant forms?

Humboldt's conception of organic form clarifies the associational logic that is involved in his physiognomy of plant collectives and indicates a logic of habitat fitness that can be methodologically identified in the broader interdisciplinary project of the new science of plant geography. How did Humboldt's conception of organic form contribute to questions concerning the geographical distribution of plants? In "Essay on the Geography of Plants," Humboldt envisioned the new science of plant geography as comprised of three disciplinary perspectives: botany, geography, and agriculture.[20] We have seen earlier how Humboldt extended the notion of botanical form as sensible shape into the logic of his physiognomy, but Humboldt's radical innovation was to set out a new topic of inquiry – the geographical distribution of plants. Humboldt clearly and explicitly thematized this principle object of inquiry, "This is the science that concerns itself with plants in their local association in the various climates."[21] In contrast to traditional botanical interests in the discovery of new species and development of a taxonomy of classes and families, Humboldt was interested in *where* plants grow geographically and, in particular, the similarities among regional plant associations. The questions concerning the geographical distribution of plants do not merely operate with an extended notion of botanical form, but a spatial notion of form that is supplied by geometry and expressed in terms of latitude and altitude. The spatial determination of geographical regions and locations can be known according to a mathematical necessity proper to geometric forms. It is in this sense that Humboldt foresaw the discipline of plant geography as "an essential part of general physics."[22] Geographical forms, for Humboldt, were spatial notions that idealize spatial location according to Euclidean planes in the tradition of Galileo and Newton. The notion of geometric form in this tradition provides a mathematical necessity to questions concerning spatial distribution. The geographical distribution of plants can be known according to a geometric notion of form measured according to geographic latitude and altitude.

The geometric notion of form provided an explanatory necessity to spatially determined geographical distributions and a backdrop from which to address the specifically botanical features of plant associations. The regional distribution of botanical associations is primarily conditioned by climate variations. Humboldt distinguished between geographical and botanical phenomenon through an appeal to climate conditions (regional temperature and atmospheric moisture). The plant forms of botanical phenomenon are conditioned by climate variations, whereas geography is spatially determined and conditioned by remote geological history. Climate variations that are determined by latitude (distance from the equator) and altitude (distance from sea level) provide the basic variable in Humboldt's logic of habitat fitness. The logic of geographical distribution of plants is organized by climate variations that find their explanatory necessity in a geometric notion of form through which latitude and altitude measurements are made possible (see Figure 2.3).

Humboldt was impressed by the similarities among plant associations and his voyage to the New World reinforced the role of climate – and, in particular, temperature – as the primary determining variable for these similarities. More than soil composition or atmospheric pressure, temperature was the determining variable of the similarities among plant associations. Temperature zones can be distinguished according to distance from the equator (latitude) and sea level (altitude) and can be graphically illustrated by isothermal maps. Humboldt correlated different temperature regions with the similarities of plant associations and illustrated how these associations changed with increasing altitude in the same way that they changed with increased distance from the equator. Isothermal maps graphically illustrated mean temperatures through isothermal lines that twist and turn according to local geographical conditions. In short,

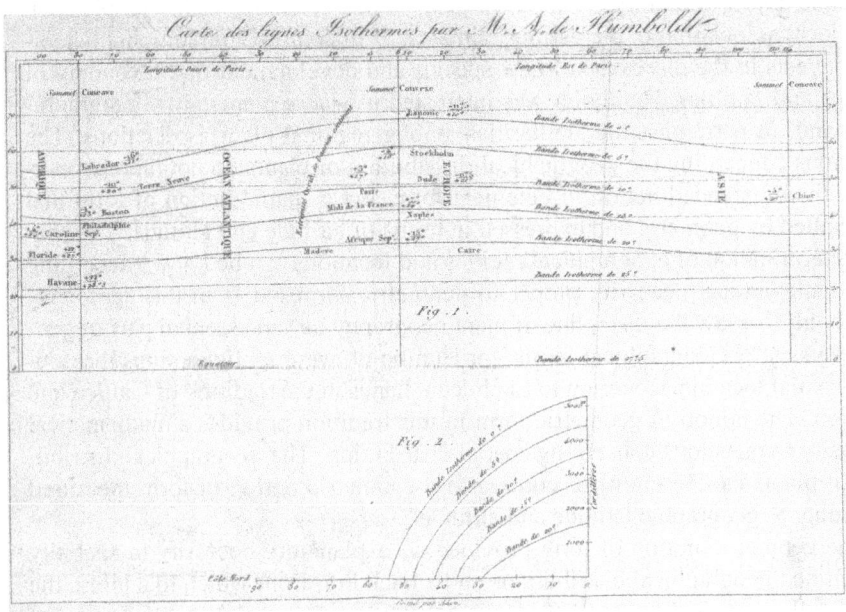

Figure 2.3 Isothermal map.

Alexander von Humboldt, "*Des lignes isothermes et de la distribution de la chaleur sur le globe,*" in *Mémories de physique et de chimie de la Société d'Arcueil* 3 (1817): 462–602.

isothermal maps represented the distribution of climate zones and provided a measurement of regional temperature variations according to which the similarities of plant associations could be correlated.

In sum, Humboldt's plant geography operated with a logic of habitat fitness that attempted to integrate various notions of form. We have seen that Humboldt's physiognomy operated with a botanical notion of form as sensible shape that was applied to the visual impressions of regional landscapes. Humboldt thus classified the forms of plant associations according to a Linnaean taxonomy of dominant species. Second, Humboldt's conception of "socially organized plants" also operated with an organic notion of form in the Kantian sense that he inherits from the German life sciences of his day. Third, Humboldt's logic of regional habitat fitness was primarily explained by means temperature variations that can be measured according to latitude and altitude and graphically illustrated through isotherms. Temperature variations primarily explain the distributional variations of plant associations and these regional temperature changes can be quantitatively measured according to a geometric notion of form. More specifically, the geographical distribution of plant forms is a spatial distribution that was modelled on the conception of absolute space in Newtonian physics. In short, Humboldt's physiognomic logic of habitat fitness illustrates a complex notion of form – 1) the sensible shapes of the visual representations of landscapes, 2) organic part-whole relations, and 3) spatial distributions according to geometric planes. Humboldt's logic of habitat fitness integrated these senses of form through an appeal to the synthetic achievements of the cognizing knower – the plant geographer, in particular. The overall unity proper to Humboldt's conception of plant forms is achieved through a synthesis of associative similarity proper to the imagination. I submit that this complex notion of plant form indicates the sense in which Humboldt's plant geography assumed an epistemological idealism that conditioned his logic of habitat fitness.

2.3 Warming's physiological logic of habitat fitness

Warming's conception of organic form operated with a different logic of habitat fitness than the physiognomic approaches inspired by Humboldt and this difference reflects different mereological suppositions regarding plant associations. The logic of habitat fitness in Warming's account of growth form was a departure from the Humboldt tradition in that it conceived of vegetative growth in terms of the chemical signatures of cellular changes involved in the nutritive process. Warming classified these growth forms according to water variations in the soil, e.g., xerophytic, mesophytic, and hydrophytic associations. Whereas Humboldt conceived of organic form in the Kantian sense and classified geographic distributions of plants according to botanical form as sensible shape, Warming's conception of "growth form" was informed by advances in 19th century plant physiology that addressed the basic question concerning the geographical distribution of plants according to a nutritive logic of habitat fitness. This nutritive conception of organic form also conceived of organisms in terms of organic parts and wholes (reciprocal purposive relations). But rather than accounting for the logic of organic form in terms of the reflective judgments that are grounded upon an *a priori* principle of purposiveness, Warming's conception of organic form accounted for the organization of plant parts and wholes in terms of the cellular assemblages involved in nutrition. This was a decisive breakthrough. This new conception of organic form was inspired by analytic and experimental chemistry which had begun to depart from the explanatory features of classical mechanics.

The difference between a botanical and nutritive conception of organic form, for Warming, had broad methodological ramifications for questions concerning the geographical distribution of plants. Warming contrasted two methodological approaches in plant geography – floristic and ecological. When the botanical notion of organic form is brought to bear on questions concerning the geographical distribution of plants, the corresponding scientific discipline would be floristic plant geography. By contrast, when a nutritive notion of growth form is brought to bear on questions in plant geography, the discipline would be ecological plant geography. Warming's *Plantesamfund* was a historical breakthrough into this new discipline:

> Oecological plant-geography has entirely different objects in view [than floristic plant-geography] – it teaches us how plants or plant-communities adjust their forms and modes of behavior to actually operating factors, such as the amounts of available water, heat, light, nutriment and so forth.[23]

Warming's nutritive logic of habitat fitness involved a methodological distinction and departure from the botanical logic of the Humboldt tradition in plant geography.

Warming's critique of Humboldt focuses, in particular, on the notion of form in Humboldt's classification,

> This is, of course, merely a superficial distinction among physiognomic and systematic types; each of these 'forms' in reality includes plants with very diverse modes of life. A purely physiognomic system is devoid of scientific significance, which is introduced only when physiognomy is founded upon physiological and oecological facts.[24]

While Humboldt's essential insight was the thematization of plant-physiognomy in relation to the landscape, his appropriation of Linnaean taxonomy in his articulation of plant forms had become dated, in Warming's view, by 19th century advances in physiology. The extension of the principles developed in the new physiology into plant geography provided an alternative functional approach to Warming's classification of growth form that promised explanatory purchase on the geographical distribution and evolution of plants.

There are at least two noteworthy experiments that illustrate the development and application of chemistry in 19th century plant physiology. The first experiment was conducted by Johannes Baptista van Helmont (1579–1644) and concerned the role of water in plant growth.[25] Helmont developed an experimental design that tested the humus theory of his day which maintained that plant growth occurred through the assimilation of humus mass – that plants grow by literally eating the soil. Helmont weighed a willow sapling, planted the sapling in potted soil, and proceeded to water the plant during a five-year period. After the duration of five years, he dislodged the willow from the potted soil and weighed each independently. The willow had significantly increased in weight, while the potted soil weighed practically the same. This indicated that the soil acts like a nutrient reservoir, but is not itself essential to plant growth. This led to the discovery that plants absorb essential mineral nutrients as inorganic ions in water. Helmont's experiment was a pioneering indication of the role of water in the absorption of soil minerals that prefigured the nutritive logic of organic form in the physiological sense.

The second and more extensive experimental approach that illustrates the introduction of chemical explanations in plant physiology was developed by Julius von Sachs (1832–1897). Sachs developed a systematic experimental approach in botany by reviving the technique of water cultures and introducing the microscopic observations as a basic kind of evidence in scientific inquiry. Sachs' development of a systematic experimental approach in botany was published in his *Lehrbuch der Botanik* (1874).[26] One of Sachs' well-known experiments concerned the dependence of the production of starch on sunlight exposure. In the experiment, Sachs compared leaves from the same plant that had been exposed to sunlight for 12 hours with leaves that did not have sunlight exposure for the same amount of time. He reduced the green pigment from the leaves and dyed the whitened leaves with iodine. The iodine reacted with the starch in the sunlight-exposed leaf, producing a blackened coloration. This reaction did not occur, however, when the iodine was applied to the sunlight-deprived leaf – it retained its white coloration. This demonstrated that only the sunlight-exposed leaves produced starch. This demonstration contributed to a nutritive conception of plant form in which plants convert light energy into chemical energy stored as starch. The carbohydrate molecules of these starches are synthesized forms of carbon dioxide and water. The water absorbed through roots and transferred through the vascular tissue shoot is incorporated in the chemical production of starches. The roots, leaves, and vascular tissue in roots are each comprised of smaller microscopic parts that are reciprocally dependent in means-end relations and contingent on external factors such as sunlight and water.

The new logic of the nutritive conception of plant form introduced by the 19th century plant physiologists was made possible by the technological advances of the microscope. This breakthrough did not occur through the field research of the new Americas – as it did for Humboldt and Darwin's breakthroughs – but in the confines of the physiologist's laboratory. The invention and refinement of the microscope enhanced the visual perception of the researcher to examine the small-scale structures of plant morphology not observable in ordinary visual perception. Analogous to the way in which Galileo's telescopes extended the reach of visual perception to Jupiter's moons, the microscope extended the reach of vision to the cellular composition of plant matter. As the physiologist peered down the barrel of the microscope, the visual perception of the researcher was provided observational access to the cellular composition of plants. This technical achievement was accompanied by a new kind of scientific evidence – microscopic observations – that introduced a new logic of parts and wholes proper to the cellular morphology and physiology of plants. It was a revolution in the history of botany that changed the basic assumptions and explanatory principles through which the plant world is known.

The revolution in plant physiology introduced a new logic of parts and wholes at work in cellular morphology and physiology of plants. This new cellular logic of nutritive forms can be particularly illustrated through Sachs' widely read textbook. The opening sentences of this book indicate this cellular conception of plant forms and is worth quoting in length.

> The substance of plants is not homogeneous, but is composed of small structures, generally indistinguishable by the naked eye. Each of these is, at least for a time, a whole complete in itself, being composed of solid, semi-solid, and fluid parts which differ in their chemical nature. These structures are termed *Cells*. For the most part, large members of them are in close contact and firmly united; and then they form a *Cellular Tissue*. But in every plant which completes its term of life there is at least one period in which certain cells or groups of cells separate at definite points from the union, and, after isolation, begin for themselves an independent course of life, as spores, pollen-grains, oospores, gemmae, etc. Like the shape and size of the whole plant, the form, structure, and size of its individual cells are subject to regular changes; and the nature of these cannot be ascertained from the study of a single phase, but only from the entire series of changes which may be called the life-history of the cell. And as such, moreover, each cell fulfills its own definite part in the economy of the plant.[27]

Cellular structures are wholes with chemically distinguishable solid, semi-solid, and fluid parts that aggregate to morphological form the members of cellular tissue. While cells and cell associations are often indistinguishable at early stages of their life-cycle, they eventually separate to form the distinct phenotypic components of the plant, e.g., roots, stems, vascular tissues, branches, leaves, and flower. The microscopic cells that comprise the small-scale heterogeneous matter of plants are comprised of "three concentrically disposed parts": 1) cell wall (firm and elastic closed envelope consisting in cellulose, 2) protoplasm (a sac enclosed by the cell wall consisting in protoplasm), and 3) nucleus (a round body embedded in the protoplasm – absent in some plants such as algae).[28] The process of cell growth occurs through an increase in cell sap that swells and expands the concentric layers of the cell as illustrated by Sachs' drawing of parenchymatous cells from the root of *Fritillaria imperialis* (Figure 2.4).

Cells are morphological wholes with parts that have regular patterns of change throughout the phases of growth. Throughout the physiological changes involved

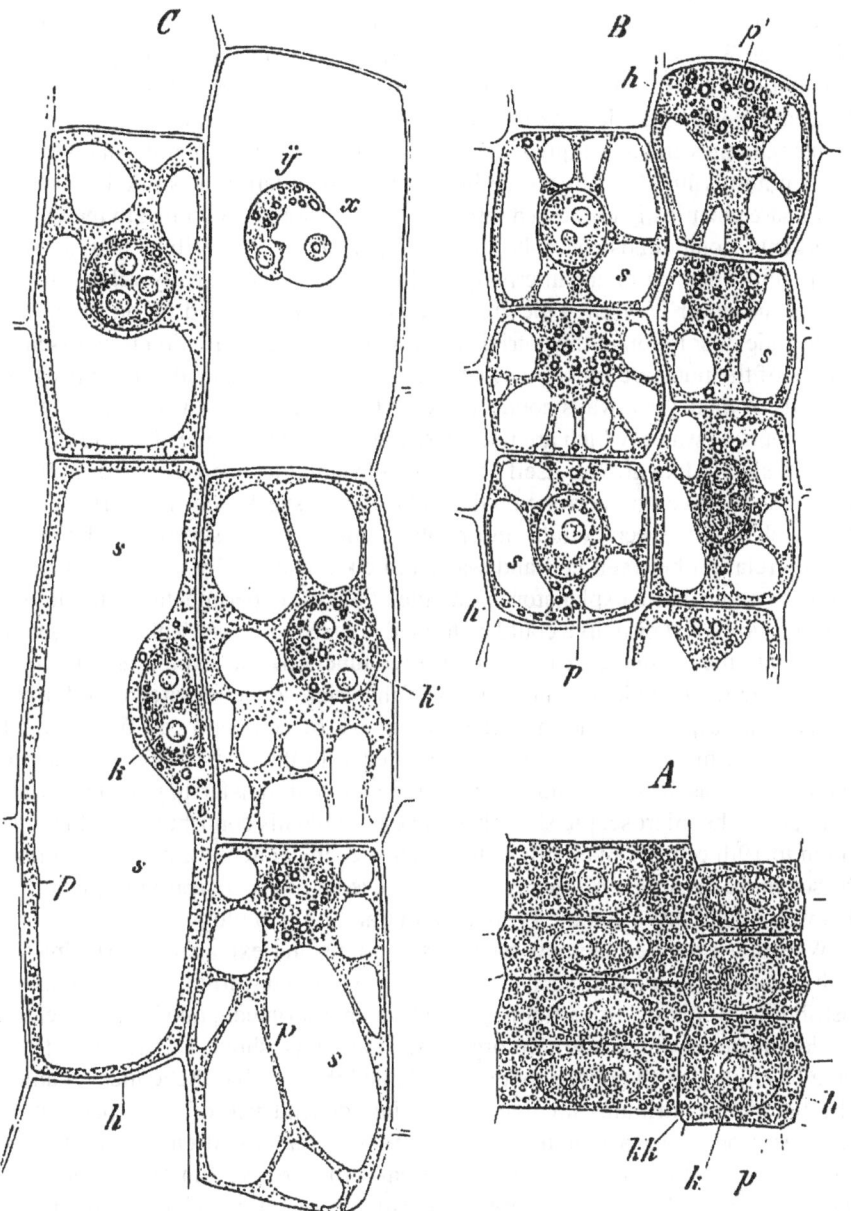

Figure 2.4 Parenchyma-cells from the central cortical layer of the root of *Frititlaria imperialis*; longitudinal sections.

Julius von Sachs, *Lehrbuch der Botanik nach dem gegenwärtigen Stand der Wissenschaft* (Leipzig: W. Engelmann, 1874); *A Textbook of Botany: Morphological and Physiological,* ed. and trans. Alfred W. Bennett (Cambridge: Cambridge University Press, 2012), 2.

in the fulfillment of nutritive functions, the organized structures of cells such as cells walls, starch grains, and protoplasms are "a combination of solid material with water."[29] The water variations involved in these molecular changes were an essential explanatory feature in Sachs' account of physiological function and he devised experimental designs that varied features of the "aqueous contents" of cellular changes and the "aqueous solution" that is absorbed in molecular growth. The variations involved in the cellular water assimilation at work in nutritive growth are chemically expressed through the provision of hydrogen to the growth of organic compounds. Morphologically speaking, the swelling and division of the molecular parts are internally unified with (not externally related to) an aquatic milieu or *Umwelt* – the "real actuality" of the water drop on the micro-scope slide or solution in the petri dish. In short, water assimilation is an essential feature of the nutritive function of cellular growth. Sachs stated, "The growth of the cells of plants is always connected with the absorption of water.... All these movements of water, which are necessarily connected with nutrition and growth, proceed slowly like growth itself."[30] Sachs' physiological conception of nutritive form thus involved a "correlational *a priori*" – to speak with Husserl – proper to the relation between cellular morphology and its water substrate. The neces-sary correlation between cell and water milieu is chemically absolute in a way that became the basic explanatory principle of Sachs' experimental methodology. This physiological principle could be termed – Sachs' *a priori*. This new essential insight into the cellular dynamics of nutritive functions was experimentally made possible by the technical achievements of the microscope to extend the horizon of visual perception to the molecular structures of cells and cell tissue. While Sachs' experimental design is widely credited with the revival of water cultures in botany, it was Sachs' methodological innovation of transferring the plant-water correlate to the microscopic slide that made possible his contribution to the revo-lution in 19th century plant physiology. This technical achievement transformed the scientific discipline of botany and introduced a new conception of plant form that was a departure from the Post-Kantian tradition in biology.

Warming extended the new basic assumptions and explanatory principles of 19th century physiologists such as Sachs to questions concerning the geographical distribution of plants. In particular, Warming appropriated a nutritive conception of plant form in which nutritive organs such as roots, shoot, and vascular tissue have physiological functions that are explained through basal chemical changes. The functional attributes involved in the interdependence of plant parts in the nutritive process are accounted for in terms of variations of chemical attributes involved in composition – the chemical basis of nutritive function. Warming's account of growth forms supposed a nutritive conception of plant growth and he self-consciously and explicitly characterized his account as an application of a physiological conception of plants to the question concerning the geographi-cal distribution of plants. Warming's physiography navigated the "borderland between the life sciences and geography" by making water variations the basis of his classifications. The water content of the soil provided a logical identity that united the nutritive variations of plant forms with the geographical forms, e.g., erosion and sediment accumulation. Warming's classifications of growth forms as xerophytic, mesophytic, and hydrophytic operated with a logic that was based on water determinations and the edaphic (Schimper) or topographical features of local habitats were conceived in terms of the local geological role of water. This logic of habitat fitness thus supposed that the manifold of plant forms and the manifold of geographical forms are necessarily united through the identity of water variations. Put differently, water variations are the identification (not association) in which biological and geographical phenomenon are united with a logical necessity. Water variations are the identity in the manifold proper to nutritive organisms and the manifold proper to the water determinations of the topography (physiography). This nutritive conception of plant forms was applied to the question concerning the geographical distribution of plants and operated with a hydrological conception of habitat fitness that was founded on (supposed by and unified with) soil moisture content as the primary explanatory feature of plant distributions. Consider the role of water variations in shaping local topographies such as soil depositions and erosion. As hills are eroded and sediment is deposited into valleys, water variations are determinative of topographical and geological land forms and provide a basis for geographical natural history. This is especially apparent in post-glacier topographies in which glacial advance and retreat are historically evidenced in river, lake, and mountain formations. The physiological ecologist makes an explanatory appeal to geological history and organizes local terrestrial and geological formations accordingly. In so doing, the ecologist is appealing to historical evidence that introduces a concern with temporality and history. Warming's critique of the conception of plant formation in the Humboldt tradition was, in part, that it was too static. Cowles developed this critique through his distinction between structural and dynamic geology, on the one hand, and structural and dynamic botany, on the other. The important point is that the physiographic ecologist takes geological formations as historical phenomenon that are accomplished as a result of a successive process. This shift to a genetic interest involved accounting for successive changes of serial formations that were historically directional – not in terms of functional sense proper to biology, but a geological sense. Ecological succession is indeed directional in this hydrological habitat fitness – not on account of the functional attributes of communities (as it is in Clements' account of communities as superorganisms), but on account of the hydrology of topographies (e.g., uplands, lowlands, ponds, valleys, ridges, rivers, and creeks). This logic of habitat fitness obtains the supposition of temporal succession in light of geological history, not the self-organization of biological collectives (ecological communities). The geological past is successively directional and when Warming and Cowles characterized the primary succession of sand dunes, for example, in holistic terms such as a "dune-embryo," they were first and foremost indicating succession of the topography, not the plant associations. In short, the physiographic logic of habitat fitness grounded the directionality of succession in topographic and geological history.

This nutritive conception of plant forms and hydrological logic of habitat fitness navigated the borderland between biology and geography by grounding the directionality of succession in natural history in the geological sense. This logical grounding of habitat changes in geological succession was different from Darwin's primary interest in the role of adaptation in the biological process of speciation in that it primarily targeted the habitat conditions themselves. Warming and the new physiological plant geographers understood and classified the organism in terms of habitat conditions that organize the distribution of plants and while this thematic focus was different than Darwin's questions about the evolutionary differences among species through natural selection, the questions were complementary.[31] The explanatory borderland of geography and biology consists of the natural history of geological formations that, in principle, could complement Darwinian natural history. While Darwin's theory of evolution by natural selection did not directly motivate the experimental designs of 19th century plant physiologists such as Sachs, the applications of the new nutritive conception of plant forms to the question concerning the geographical distribution of plants directly addressed Darwin's account of the reproductive fitness involved in natural selection.

The application of a nutritive conception of plant forms to investigations concerning plant distributions was considered methodologically rogue in many ways. The experimental constraints of the laboratory provided the isolation of variables that was lost in the fieldwork of plant geography. The complexity and contingency of variables involved in questions concerning plant distribution de-motivated serious scientific attention. For 19th century plant physiologists, the laboratory was the site in which investigations concerning the chemical expression of plant nutrition were conducted and the methodological constraints of the experimental design of laboratory work provided the insight into the physical laws that govern plant growth. Questions concerning plant distribution that arose in plant geography relied on an outdated floristic classification that advances in plant physiology had left behind and were largely uninteresting in light of the physiological supplement of botany. Warming's "ecological plant geography" was a breakthrough that illustrated the relevance of the new nutritive conception of plants for questions concerning plant distributions and its potential implications for the methodological resources in evolutionary biology. More to the point, the contingencies involved in fieldwork can be quantitatively measured through population statistics. Thomas Malthus' innovative application of probability theory to London mortality rates from 1780–1810 demonstrated the effectiveness of differential equations for large-scale data sets.[32] The Malthusian scarcity curve demonstrated a positive correlation between population abundance (data sets from obituaries and hospital birth records) and agricultural yield of the surrounding region.

Malthus' methodological innovation impressed Darwin and his followers and by 1930 Robert Fischer had mathematically expressed the basic explanatory principle of evolutionary biology in "The Fundamental Theorem of Natural Selection."[33] The use of probability theory to mathematically address large data sets would soon spread to other spheres of inquiry, including questions concerning the geographical distribution of plants. While 19th century ecologists such as Warming might have been considered methodologically rogue from the perspective of the experimental designs of the plant physiologists, they were methodologically confident in the resources of probability theory to establish correlations among population abundance of plant species with various environmental constraints. To be sure, questions concerning plant distribution are different than physiological questions concerning the chemical basis of nutrition. But the application of the new nutritive conception of form, in principle, had implications for explanations concerning plant distributions. In particular, the mineral content of the soil had direct relevance for the nutritive processes that condition population statistics and provided a basic datum for mathematical measurement.

Warming's application of a nutritive conception of organic form to the question concerning the geographical distribution of plants was methodologically rogue from the perspective of the plant physiologists working in the laboratory to isolate variables in order to discover mechanistic explanations with an efficient causal necessity. Questions concerning the geographical distribution of plants involved an abandonment of the experimental method that establishes the necessary relations among nutritive variations and introduced a wide array of uncontrolled variables. Warming and the early physiological plant ecologists, however, were confident in the mathematical resources stemming from probability theory to address the stochasticity inherent in questions concerning plant distribution in way that was consistent with Darwin's insight into and Fischer's formalization of the theory of natural selection. These pioneering "physiological ecologists" or "ecological plant geographers" or "outdoor physiologists" were not completely unmoored from an ability to chemically identify and mathematically measure the processes involved in plant nutrition. The application of population statistics to questions concerning plant distributions, in principle, could provide the methodological resources to explain the biological processes of how plant associations change (succession). In short, while Warming and the other early physiological plant ecologists were departing from the experimental framework of the laboratory and the strong explanatory necessity that it afforded, they were methodologically bolstered by the successful application of population statistics in evolutionary theory. Accompanying this mixed methodology was a combination of a nutritive conception of organic form and an evolutionary conception of organic function in terms of reproductive fitness.

Ernst Haeckel's early vision of the science of ecology in many ways anticipated the application of physiological notions of organic form in plant geography later in the 19th century. Haeckel's *Generelle Morphologie der Organismen* (1866) was primarily an investigation of morphology from a Darwinian approach to adaptation, but includes a brief section in which Haeckel coined the term *"Ökologie"* as a new domain of scientific inquiry with the relation between an organism and its environment as its principle object. He recognized that both the organic and inorganic conditions of the relations of organisms and environments characterized the organic conditions of ecological investigation in expressly physiological terms. More specifically, Haeckel called for a two-sided approach in physiology – inquiry not only into the nutritive processes internal to the organic parts of individuals and species, but the nutritive processes involved in the organic conditions of the relation of organisms to their external environment. Haeckel's definition of ecology and challenge to physiology to ask broader questions is worth quoting at length:

> By ecology, we mean the whole science of relations of the organism to the environment, including, in the broad sense, all the 'conditions of existence.' These are partly organic, partly inorganic in nature; both, as we have shown, are of the greatest significance for the form of organisms, for they force them to become adapted. Among the inorganic conditions of existence to which every organism must adapt itself belong, first of all, the physical and chemical properties of its habitat, the climate (light, warmth, atmospheric conditions of humidity and electricity), the inorganic nutrients, nature of the water and of the soil, etc. As organic conditions of existence we consider the entire relations of the organism to all other organisms with which it comes into contact, and of which most contribute either to its advantage or its harm. Each organism has among the other organisms its friends and its enemies, those which favor its existence and those which harm it. The organisms which serve as organic foodstuff for others or which live upon them as parasites also belong in this category of organic conditions of existence. In our discussion of the theory of selection we have shown what enormous importance all these adaptive relations have for the entire formation of organisms, and specially how the organic conditions of existence exert a much more profound transforming action on organisms than do the inorganic. The extraordinary significance of these relations does not correspond in the least to their scientific treatment, however. So far physiology, [the science] to which this belongs, has, in the most one side fashion, almost exclusively investigated the conserving functions of organisms (preservation of the individual and the species, nutrition, and reproduction), and among the functions of relationship [it has investigated] merely those which are produced by the relations of single parts of the organism to each other and to the whole. On the other hand, physiology has largely neglected the relations of the organisms to the environment, the place each organism takes in the household of nature, in the economy of all nature, and has abandoned the gathering of the relevant facts to an uncritical 'natural history,' without making an attempt to explain them mechanistically. This great gap in physiology will now be completely filled by the theory of selection and the theory of evolution which results directly from it.[34]

Perhaps the most noteworthy aspect of this extended quotation concerns Haeckel's conception of the relation of physiology to evolutionary biology. Haeckel's condensed argument runs as follows. The habitat conditions that structure an organism's adaptation are both inorganic and organic. The inorganic conditions are the physical and chemical properties of the habitat. Physiology serves as the study of organic conditions that includes a study of the internal organic processes (preservation, nutrition, and reproduction of individuals and species), on the one hand, and the external processes among collectives of inter-specific organisms, on the other. The later inter-specific organic conditions have been neglected by physiology and natural selection provides physiology with an explanatory principle to address this oversight. In short, Haeckel foresaw the potential extension of physiological accounts of organisms to inter-specific interactions among organic associations and, at least in this abstract methodological sense, the late 19th century physiological ecologists such as Warming were fulfilling this principle of Haeckel's vision.

Warming's emphasis, however, was on the inorganic conditions of plant habitats. I have argued that his mereological assumptions and methodological principles can be distinguished from the physiognomy of the Humboldt tradition by highlighting the historical influence of 19th century plant physiology and evolutionary biology on Warming's approach. Warming's approach to plant geography could also be distinguished from Humboldt's through an identification of his logic of habitat fitness. There are several noteworthy features of Warming's nutritive conception of plant forms and logic of habitat fitness that highlight the difference with the Humboldt tradition.

1) Geographical succession is organized by water variations involved in glaciation, erosion, and sediment accumulation. Water variations organize these topographical processes with a directionality toward a baseline.
2) Water variations organize the nutritive processes involved in plant distributions. Plant associations are thus classified according to water tolerances, e.g., xerophytes, mesophytes, and hydrophytes.[35]
3) Water variations are the identity in both the manifold of geographical variations and plant distributions.
4) Plants are geographically distributed according to "growth forms" – the structural and behavioral characteristics of plants growing in a similar habitat.
5) Plant formations are defined as a community of species which have become characteristically associated by edaphic and climatic features of a habitat.[36]
6) The basic unit of a plant association is the individual plant.[37]

An initial point of entry into Warming's nutritive logic of habitat fitness can be provided by Warming's notion of "growth form." Warming generally characterized a growth form as follows,

It reveals itself especially in the habit, and in the form and duration, of the nutritive organs (in the structure of the foliage-leaf and of the whole vegetative shoot, in the duration of life of the individual, and so forth), but shows to a less extent in the reproductive organs. This subject leads us into deep morphological, anatomical, and physiological investigations; it is very difficult, yet very alluring; but only in a few cases can its problems be satisfactorily solved at the present time.[38]

A growth form concerns the relationship between nutritive features of the organism and its habitat and classifies vegetation according to the principles of plant physiology rather than the floristic taxonomy developed in Linnaean botany. Warming distinguished his notion of growth form from the "vegetative form" introduced by August Grisebach and employed in systematic botany to classify the diversity of floral structures and methods of pollination. By contrast, a growth form concerns the nutritive function of plant cells, tissues, and organs through the clarification of specialized chemical functions. As we have seen in the previous investigation, roots and rhizoids, for example, acquire minerals from the soil and anchor the plant. Leaves catch light and contain chlorophyll – a chemical compound that interacts with light and enables plants to produce their own nutrients. The vascular tissues in plant shoots (e.g., stem, trunk, and so on) transport water and minerals from the roots to the leaves. The cellular structure and chemical changes of these roots, leaves, and vascular tissues are different in a way that clarifies the distinct function of these organs in the nutritive process. Water, in particular, is a chemically expressed compound (H_2O) whose oxygen molecule is incorporated into the process of photosynthesis. This represents the chemicalization of nutritive function, which is to say, nutritive function is explained by its underlying chemical conditions and demonstrates the chemical interdependence of nutritive organs. This is a decisive logical difference with Humboldt's botanical conception of organic form in that it explains the organic functions of nutrition and growth in terms of underlying chemical variations rather than a reflective principle of purposiveness in the Kantian sense. In short, Warming's conception of a growth form operated with a nutritive conception of organic form that organized his logic of habitat fitness.

Second, Warming's nutritive conception of growth form organized his logic of habitat fitness through local water variations rather than regional climate variations. At first glance, the difference between these two manifolds (water and climate variations) could be considered through a distinction between local and regional scales. Climate variations condition plant associations at a regional level in that temperature and rainfall are broadly distributed due to large scale conditions such as seasonal planetary tilt and daily rotations, ocean currents, and so on. By contrast, water variations in the soil track the edaphic or local distributions of plant associations that are conditioned by geographic and geological factors. While this local-regional scale distinction is relevant in distinguishing a physiological and physiognomic logic of habitat fitness, the two scales are not mutually exclusive. Both determinate conditions can be relevant without a logical conflict through a focus on particular habitats. The logical issue concerning habitats involves the principle characteristics and conditions that are determinative in the geographical distribution of plants.

Third, as we have seen in Cowles' account of primary dune succession in the previous investigation, the same water variations that condition plant distribution also condition the water variations involved in topographical changes, e.g., erosion, sediment accumulation, and so on. Cowles followed Warming in this emphasis on water variations as the identity in both manifolds of topographical changes and nutritive processes. The water variations that determine the topographical changes in processes of erosion have an overall base leveling effect – hills are worn away and valleys are filled – and provide a directionality to these topographical changes. The directionality involved in Warming's account of plant distribution is not derived from an organic notion of function of individual organisms that was misapplied to plant collectives. Rather, this directionality is derived from the tendency of topographical base leveling involved in, for example, erosion and sediment accumulation. Warming navigated the "borderland between the biology and geography" by grounding his account of plant succession in an account of geographical succession. More specifically, geographical (inorganic) succession has a founding relation with the organic attributes of plant succession. This means that organic plant succession presupposes and is organized by inorganic geographical succession. This is the sense in which Warming characterized his methodological approach through the neologism "physiography."

Fourth, the same water variations that organize the topographical features of habitats also organize the distribution of growth forms into various plant associations. Water variations were the basic characteristics of Warming's classification of plant communities (characteristic or typical plant associations). Warming stated,

> The grouping of the classes of communities here adopted is based in the first place upon the plant's dependence upon and relation to water. Pindar's aphorism, 'water is best of all,' is wholly true of plant-life; water is the condition of life that exercises the greatest influence in bringing into being external and internal differences among plants; it is likewise water that plays the leading part in determining the creation of plant-communities and their distribution over the soil. It is quite true that the special attributes of a habitat result from the most diverse factors, edaphic and climatic... But it is beyond doubt that water occupied the foremost position as a factor bringing about the greatest distinctions in vegetation and structure.[39]

"Water is best of all" in Warming's logic of habitat fitness in the sense that water tolerances are the principle characteristic of the classification of plant communities, e.g., xerophytic, mesophytic, and hydrophytic. The logic of habitat fitness in this taxonomy is organized according to an identification of the same variable (surface groundwater) in both the geographic (inorganic) manifolds of change and the nutritive (organic) manifolds of change involved in plant succession. This is logically significant for a variety of reasons that can be highlighted through a contrast with Humboldt's logic. Recall Humboldt's distinction between geographical and vegetative phenomenon that are united in a landscape and are distinguished by climate conditions (temperature and atmospheric moisture). The plant forms of vegetative phenomenon are conditioned by climate variations, whereas geography is conditioned by geological history (not proximate climate variations). There are two manifolds of change – geographical (inorganic) and vegetative (organic) – that are distinguished by Humboldt according to climate (and temperature, in particular) variations. Whereas the vegetation of a regional landscape is primarily conditioned by temperature variations, temperature does not significantly condition the geographical features of a landscape. Humboldt thus had two manifolds of change with two respective unities of identification (temperature variations and geographical history). His pioneering attempt to formulate the science of plant geography was dualistic in the sense that it attempted to synthesize two manifolds that were conceived in an external relation. By contrast, Warming's habitat logic internally unified the manifold of variations in geographical (inorganic) succession with the manifold of variations in nutritive (organic) succession through the identification of soil water composition as the principle conditioning characteristic. The manifolds of variation (topological and nutritive) in this logic are thematized according to one and the same identity – soil moisture content – that internally unifies both nutritive growth and topological variation.

2.4 Conclusion

This investigation has identified and preliminarily clarified the sense in which Warming's nutritive conception of plant form organized his physiographic logic of habitat fitness in a way that is different from Humboldt's fundamental assumptions and explanatory principles. There are two kinds of claims in this argument – historical and logical. My historical claim is that 19th century plant geography provides at least two traditions that can be contrasted through their respective accounts of organic form and logic of habitat fitness. Humboldt's botanical classification of the physiognomic logic of habitat fitness took its explanatory departure from the visual impressions of regional landscapes and accounted for the unity of plant associations in terms of a Kantian conception of organic form. By contrast, Warming's nutritive conception of plant forms was primarily influenced by discoveries in 19th century plant physiology that explained plant nutrition through underlying chemical conditions and demonstrated the chemical interdependence of nutritive organs. This is a decisive logical difference with Humboldt's botanical conception of organic form in that it explains the organic functions of nutrition and growth in terms of underlying chemical variations rather than a reflective principle of purposiveness in the Kantian sense. My logical claim is that the differences in these conceptions of organic form imply differences in the logical accounts of habitat fitness provided by Humboldt's physiognomy and Warming's physiography. We have seen that these respective logics of habitat fitness bring different fundamental assumptions and explanatory principles in 19th century plant geography.

This investigation has also highlighted the sense in which Humboldt's physiognomic logic of habitat fitness assumed a complex notion of plant form. Warming's distinction between Humboldt's "floristic plant geography" and his "ecological plant geography" utilized a critique of the botanical conception of plant form as sensible shape. In conclusion, I would like to problematize the sense in which Humboldt's logic of habitat fitness supposed an organic conception of plant form. We have seen that Humboldt's pioneering conception of socially organized plants generalized a notion of organic form of individual plants to plant associations as organized and organizing collectives. I submit that this generalization of part-whole relations proper to individual organisms to account for the holistic attributes of plant associations is made possible through an epistemological idealism that grounds the associative unity of organic parts and wholes in the synthetic achievements of the cognizing knower. Humboldt's notion of plant form in the organic sense accounted for the unity of organisms by making an appeal to a different kind of explanatory principle – an *a priori* principle of purposiveness – that is at work in aesthetic and teleological judgments in the Kantian sense. The following investigation is an immanent critique of this conception of organic form and the epistemological idealism that makes it possible. In a later investigation (Chapter Five), I problematize the sense in which Humboldt's logic of habitat fitness assumed a geometric notion of form in its conception of spatial distribution in the Newtonian sense.

My theoretical motivations for problematizing the notion of organic form in the Kantian sense are both historical and logical. Historically speaking, the genealogies of early 20th century American plant ecology and 19th century plant geography illustrate a persistent philosophical tension concerning the logic of habitat associations and conceptions of plant formations. In other words, the search for an ecological notion of form in these genealogies illustrates a split between the fundamental assumptions and explanatory principles of the epistemological idealism of Humboldt and Clements, on the one hand, and the epistemological realism of Warming and Cowles, on the other. In short, my historical motivation in these investigations involves an attempt to philosophically clarify this logical and epistemological tension within the history of the ecological sciences.

Second, the historical motivations that are involved in Humboldt and Kant's conception of organic form gain significance in light of the persistent logical difficulties involved in contemporary community ecology that appeals to multi-level explanations. Debates in community ecology concerning unity of analysis problems, reductionistic and holistic explanations, and functional analysis of ecological collectives remain philosophical puzzles with a plurality of methodological approaches. I submit that Kant's shadow still looms large in these contemporary debates, particularly with regard to the fundamental assumptions and explanatory principles of contemporary research that attributes self-organization to ecological associations. My logical motivation in the subsequent investigations is thus a departure from the genealogical orientation of the previous investigations that prioritized a historical interest in the diagnosis of the unique philosophical issues involved in the ecological sciences. The dialogues with Kant, Husserl, and Newton in the following investigations are primarily motivated by a logical interest in the ecological things themselves. More specifically, the following investigations into Kant's conception of organic form and the proximate explanations in biology, the logic of fitness in Husserl's formal theory of part-whole relations, and the distinction between absolute space in the Newtonian sense and geographical places – each of these investigations are historical engagements of philosophically persistent issues in the history of the ecological sciences. My logical motivation in these investigations involves the development of my own position regarding these issues. Historically speaking, the positions I develop in the subsequent investigations are largely inspired by the phenomenological movement of the early 20th century. I employ resources from this phenomenological tradition, especially Husserl's theory of intentionality, logic of part-whole relations, and distinction between formal and regional ontology in an attempt to strengthen epistemological realist approaches in population and community ecology, e.g., Warming and Cowles' physiographic ecology. While my logical interest does not directly engage contemporary debates in population and community ecology of the late 20th century, it nevertheless radically pushes through to the ecological things themselves and, in particular, to the forms proper to habitat associations.

Notes

1 Alexander von Humboldt, "Ideas for a Physiognomy of Plants," in *Views of Nature*, eds. Stephen T. Jackson and Laura Dassow Walls, trans. Mark W. Person (Chicago: University of Chicago Press, 2014): 155–242; *Ansichten der Natur* (Stuttgart and Tübingen: Cotta, 1849); Oscar Drude, "The Position of Ecology in Modern Science," in *Congress of Arts and Science: Universal Exposition, St. Louis, 1904*, ed. Howard J. Rogers, trans. Jane Patten (Boston: Houghton, Mifflin and Company, 1906): 177–190; Eugene Warming, *Oecology of Plants: An Introduction to the Study of Plant-Communities* (Oxford: Clarendon, 1909): 2–13.
2 Henry Chandler Cowles, "The Ecological Relations of the Vegetation on the Sand Dunes of Lake Michigan. Part I – Geographical Relations of the Dune Flores," *Botanical Gazette* 27, no. 2 (1899): 95.
3 Ronald Tobey, *Saving the Prairies: The Life Cycle of the Founding School of American Plant Ecology*, 1895–1955 (Berkeley: University of California Press, 1981); Frank Benjamin Golley, *A History of the Ecosystem Concept in Ecology: More Than the Sum of the Parts* (New Haven: Yale University Press, 1993).
4 Humboldt, "Ideas for a Physiognomy of Plants," 156.
5 Ibid., 160.
6 Ibid., 161.
7 Ibid., 158.
8 Ibid., 160.
9 Humboldt, "Ideas for a Physiognomy of Plants," 158.
10 Kurt Koffka, *Principles of Gestalt Psychology* (London: Routledge, 1955).
11 "Nature considered rationally, that is to say, submitted to the process of thought, is a unity in diversity of phenomena; a harmony, blending together all created things, however dissimilar in form and attributes; one great whole…animated by the breadth of life. The most important result of a rational inquiry into nature is, therefore, to establish the unity and harmony of this stupendous mass of force and matter, to determine with impartial justice what is due to the discoveries of the past and to those of the present, and to analyze the individual parts of natural phenomena without succumbing beneath the weight of the whole." Alexander von Humboldt, *Cosmos: A Sketch of the Physical Description of the Universe, Volume 1*, trans. E.C. Otté (Baltimore and London: Johns Hopkins University Press, 1997): 24. First published as *Kosmos: Entwurf einer physischen Weltbeschreibung, Erfter Band*, ed. Traugott Bromme (Stuttgart and Tübingen: Cotta, 1845). See also Humboldt, "Ideas for a Physiognomy of Plants," 159.
12 Ibid., 239.
13 See Antonio Carlos Vitte, Dias da Silveira, and Roberison Wittgenstein, "*Kant, Goethe E Alexander Humboldt: Estética E Paisagem Na Génese Da Geografia Física Moderna*," *ACTA Geográfica, Boa Vista*, 4, no. 8 (2010): 7–14.
14 Humboldt, "Ideas for a Physiognomy of Plants," 239.

15 Alexander von Humboldt, *Florae Fribergensis specimen, plantas cryptogamicas prae-sertim subterraneas exhibens* (Berolini: H. A. Rottman, 1793): 133. For a more extensive discussion, see Robert J. Richards, *The Romantic Conception of Life: Science and Philosophy in the Age of Goethe* (Chicago: University of Chicago Press, 2002): 313–321.

16 Alexander von Humboldt, *Versuche über die gereizte Muskel- und Nervenfaser nebst Vermutungen über den chemischen Process des Lebens in der Tier- und Pflanzenwelt*, Vol. 2 (Berlin: Heinrich August Rottmann, 1797–99).

17 Luigi Galvani, *Commentary on the Effects of Electricity on Muscular Motion*, ed. Bernard Cohen, trans. Margaret Foley (Norwalk, Conn.: Burndy Library, 1953).

18 Alessandro Volta, *De VI Attractiva Ignis Electrici, Ac Phaenomenis Inde Pendentibus Alexandri Voltae, Dissertatio Epistolaris* (Novocomi: Octavii Staurenghi, 1769).

19 Alexander von Humboldt, *Versuche über die gereizte Muskel- und Nervenfaser nebst Vermutungen über den chemischen Process des Lebens in der Tier- und Pflanzenwelt*, Vol. 2 (Berlin: Heinrich August Rottmann, 1797–99): 433–434. Translation from Richards, *The Romantic Conception of Life*, 320.

20 Humboldt, *Essay on the Geography of Plants*, 64–75.

21 Ibid., 64.

22 Ibid., 64.

23 Eugene Warming, *Oecology of Plants: An Introduction to the Study of Plant-Communities* (Oxford: Clarendon, 1909): 2.

24 Ibid., 4.

25 Johannes Baptista van Helmont, *Ortus Medicinae. Id Est Initia Physicae Inaudita. Progressus Medicinae Novus* (Amsterdam: L. Elzevir, 1652). This experiment is simulated in Matt Barrett's production of *Botany: A Blooming History* (London: BBC Documentaries, 2014).

26 Julius von Sachs, *Lehrbuch der Botanik nach dem gegenwärtigen Stand der Wissenschaft* (Leipzig: W. Engelmann, 1874); *Textbook of Botany*, 2nd ed. (Oxford: Clarendon Press, 1882).

27 Sachs, *Textbook of Botany*, 1.

28 Ibid., 2.

29 Ibid., 663.

30 Ibid., 674.

31 Gregory J. Cooper, *The Science of the Struggle for Existence: On the Foundations of Ecology* (Cambridge: Cambridge University Press, 2003): 35–37. Cooper highlights three differences between Warming's physiological ecology and Darwin's evolutionary biology: 1) their respective approaches to adaptation, 2) Darwin's focus on the individual organism and Warming's focus on plant associations, and 3) a contrast between historical and mechanistic forms of explanations. Cooper distinguishes these fields of inquiry according to proximate and remote explanations to adaptation, "As ecology began to take on more of the character of experimental field physiology, this link began to weaken; the investigations tended to be directed at proximate questions about organism/environment interaction" (55). In contrast, my own view is that the basic difference simply concerns the basic questions and objects of inquiry. While Darwin was primarily focused on the differences among species (Why are there so many difference species of plants and animals?), Warming was primarily focused on the geographical distribution of plants (How are plants geographically distributed?). These are related and complementary questions, but they are different. Cooper's appropriation of the distinction between proximate-remote explanations nevertheless has purchase on distinguishing the basic questions of these two regions of inquiry. The Darwinian theory of evolution by natural selection addresses the diversity question in terms of the remote historical origins of species differentiation. Generally speaking, it explains differences through a search for remote origins. By contrast, the question concerning

the geographical distribution of plants is addressed by Warming in terms of proximate nutritive conditions, e.g., light, water, soil composition, climate, population and community interactions, and so on. Generally speaking, it explains plant distribution in terms of proximate habitat factors that condition the plant associations of particular geographical locations. Warming was not, however, Lamarckian. That is, he did not attempt to account for plant adaptation through proximate explanations. Rather, he was concerned with the geographical conditions of distribution – the habitat fitness – that set the parameters for plant adaptation. This attempt did indeed incorporate an appeal to historical explanations, but these historical explanations were fundamentally physiographic (surface geological and topological), not biological.

32 Thomas Robert Malthus, *An Essay on the Principle of Population* (London: J. Johnson, 1798).
33 Robert Fisher, *The Genetical Theory of Natural Selection* (Oxford: Clarendon Press, 1930).
34 Ernst Haeckel, *Generelle Morphologie der Organismen: Allgemeine Grundzüge der organischen Formen-Wissenschaft, mechanisch begründet durch die von Charles Darwin reformirte Descendenz-Theorie* (Berlin: Reimer, 1866). Translation from Robert C. Stauffer, "Haeckel, Darwin, and Ecology," *Quarterly Review of Biology* 32: 140–141.
35 Warming, *Oecology of Plants*, 96.
36 Ibid., 140.
37 Ibid., 91.
38 Ibid., 3.
39 Ibid., 96.

Bibliography

Cooper, Gregory J. *The Science of the Struggle for Existence: On the Foundations of Ecology.* Cambridge: Cambridge University Press, 2003.

Cowles, Henry Chandler. "The Ecological Relations of the Vegetation on the Sand Dunes of Lake Michigan. Part I – Geographical Relations of the Dune Flores." *Botanical Gazette* 27, no. 2 (1899): 95–117.

Drude, Oscar. "The Position of Ecology in Modern Science." In *Congress of Arts and Science: Universal Exposition, St. Louis, 1904*, edited by Howard J. Rogers, translated by Jane Patten, 177–190. Boston: Houghton, Mifflin and Company, 1906.

Fisher, Robert. *The Genetical Theory of Natural Selection.* Oxford: Clarendon Press, 1930.

Galvani, Luigi. *Commentary on the Effects of Electricity on Muscular Motion*, edited by I. Bernard Cohn, translated by Margaret Foley. Norwalk, CT: Burndy Library, 1953.

Golley, Frank Benjamin. *A History of the Ecosystem Concept in Ecology: More Than the Sum of the Parts.* New Haven: Yale University Press, 1993.

Haeckel, Ernst. *Generelle Morphologie der Organismen: Allgemeine Grundzüge der organischen Formen-Wissenschaft, mechanisch begründet durch die von Charles Darwin reformirte Descendenz-Theorie.* Berlin: Reimer, 1866.

Helmont, Johannes Baptista van. *Ortus Medicinae. Id Est Initia Physicae Inaudita. Progressus Medicinae Novus.* Amsterdam: L. Elzevir, 1652.

Humboldt, Alexander von. *Florae Fribergensis Specimen, Plantas Cryptogamicas Praesertim Subterraneas Exhibens.* Berolini: H. A. Rottman, 1793.

Humboldt, Alexander von. *Versuche über die gereizte Muskel- und Nervenfaser nebst Vermutungen über den chemischen Process des Lebens in der Tier- und Pflanzenwelt,* Vol. 2. Berlin: Heinrich August Rottmann, 1797–1799.

Humboldt, Alexander von. *"Des lignes isothermes et de la distribution de la chaleur sur le globe."* *Mémories de physique et de chimie de la Société d'Arcueil* 3 (1817): 462–602.

Humboldt, Alexander von. *Kosmos: Entwurf einer physischen Weltbeschreibung, Erſter Band*, edited by Traugott Bromme. Stuttgart and Tübingen: Cotta, 1845.

Humboldt, Alexander von. *Ansichten der Natur*. Stuttgart and Tübingen: Cotta, 1849.

Humboldt, Alexander von. *Cosmos: A Sketch of the Physical Description of the Universe, Volume 1*, translated by E. C. Otté. Baltimore and London: Johns Hopkins University Press, 1997.

Humboldt, Alexander von. *Essay on the Geography of Plants*, edited by Stephen T. Jackson, translated by Sylvie Romanowski. Chicago: University of Chicago Press, 2009.

Humboldt, Alexander von. "Ideas for a Physiognomy of Plants." In *Views of Nature*, edited by Stephen T. Jackson and Laura Dassow Walls, translated by Mark W. Person, 155–242. Chicago: University of Chicago Press, 2014.

Koffka, Kurt. *Principles of Gestalt Psychology*. London: Routledge, 1955.

Malthus, Robert Thomas. *An Essay on the Principle of Population*. London: J. Johnson, 1798.

Richards, Robert J. *The Romantic Conception of Life: Science and Philosophy in the Age of Goethe*. Chicago: University of Chicago Press, 2002.

Sachs, Julius von. *Lehrbuch der Botanik nach dem gegenwärtigen Stand der Wissenschaft*. Leipzig: W. Engelmann, 1874.

Sachs, Julius von. *A Textbook of Botany: Morphological and Physiological*, edited and translated by Alfred W. Bennett. Cambridge: Cambridge University Press, 2012.

Stauffer, Robert C. "Haeckel, Darwin, and Ecology." *Quarterly Review of Biology* 32, no. 2 (1957): 140–141.

Tobey, Ronald. *Saving the Prairies: The Life Cycle of the Founding School of American Plant Ecology, 1895–1955*. Berkeley: University of California Press, 1981.

Vitte, Carlos Antonio, da Silveira, Dias, and Wittgenstein, Roberison. "Kant, Goethe E Alexander Humboldt: Estética E Paisagem Na Génese *Da* Geografia Física Moderna." *ACTA Geográfica, Boa Vista* 4, no. 8 (2010): 7–14.

Volta, Alessandro. *De VI Attractiva Ignis Electrici, Ac Phaenomenis Inde Pendentibus Alexandri Voltae, Dissertatio Epistolaris*. Novocomi: Octavii Staurenghi, 1769.

Warming, Eugene. *Oecology of Plants: An Introduction to the Study of Plant-Communities*. Oxford: Clarendon, 1909.

3 Kant's account of organic form

A phenomenological critique

3.1 Introduction

This investigation focuses on the logic of organic forms that are involved in functional descriptions of individuals. My dialogue with Immanuel Kant in this investigation is an attempt to identify and clarify a fundamental epistemological error of biological (and ecological) idealism. I argue that Kant's conception of the unity of organic forms operates with an epistemological formalism or one-sidedness and that a stronger epistemological realism involved in a phenomenological approach can provide a better alternative. I argue that this phenomenological alternative is uniquely suited to philosophically complement the notion of "growth form" in Eugene Warming's pioneering approach to physiographic ecology.

The focus of this investigation is more fully motivated by the philosophical issues involved in the logic of organic forms that is inherently inclusive of the sense-making involved in various habitat relations. This motivation is a departure from historical considerations and instead is oriented toward the persistent philosophical questions concerning biological and ecological forms that remain even in a contemporary context, e.g., the notion of biological form that is implicit in Ernst Mayr's articulation of proximate explanations. What are the logical suppositions regarding organisms that are involved in proximate explanations? I argue later that Kant's one-sided answer to this question is inadequate and, more specifically, that his account of synthesized manifolds underdetermines the unity proper to organisms. Moreover, I argue that a better answer to this question concerns the way in which organisms are given as identities-in-a-manifold that are organized by sense-making processes, that is, as biological objects that are unified and organized in their own right as individuals. In short, my claim is that functional descriptions in biology need some kind of an account of biological form and I argue that biological forms are identities-in-a-manifold that are given with a layer of functional sense that can be characterized with the necessity involved in a phenomenological logic of habitat fitness.

This investigation is comprised of three related theses. First, this chapter involves a reconstruction of Kant's account of organic form. I argue that Kant's account of organic form has at least three essential features: A) organisms have interdependent parts and wholes, B) the interdependence of part and wholes proper to organisms are contingent rather than necessary, and C) organic parts and wholes can be properly characterized in reflective judgments as reciprocal means-end or purposive relations. This argument could be considered my reconstructive thesis. Kant's account of organic form in the second part of the *Critique of the Power of Judgment* relies on the epistemological framework of the *Critique of Pure Reason*, particularly Kant's broader account of synthesized manifolds and the role of schemata in Kant's distinction between determinate and reflective judgments. My reconstruction of Kant's overall account is an attempt to get Kant right, so to speak, and I realize that a full demonstration of this thesis would require its own sustained treatment that would be beyond the confines of this investigation.[1] Nevertheless, I submit that Kant's conception of organic form has at least three identifiable essential features and that his attempt to ground the necessity involved in the principle of purposiveness clarifies these features of his account.

Second, I argue that Kant's search for an *a priori* principle to ground purposive judgments underdetermines the kind of unities that are involved in organic forms themselves. Kant's account is one-sided – it grounds the necessary unity of organic forms in the synthetic achievements of the cognitive subject in an asymmetrical relation and thereby underdetermines the kinds of unity proper to the organic forms of individuals themselves. This argument could be considered my critical thesis.[2]

Third, the philosophical motivation for this chapter concerns the role of organic forms in functional descriptions. I argue that proximate explanations in functional biology are indispensable and legitimate features of contemporary biological inquiry. Proximate explanations operate with certain conceptions of organic form, moreover, whether or not they are methodologically clarified. Functional descriptions in biology suppose some kind of logical account of organic form, that is, they suppose a relation of parts and wholes that are organized in various ways that can and should be approached with various experimental and mathematical methodologies. Within this broader pluralistic context, my phenomenological approach to the logic of organic form is focused on the organization of part-part and part-whole relations that are presupposed by the functional sense of an individual organism's habitat fitness. My claim is that the sense-making of the phenotypic individual in its habitat is the primary material content of biological forms. This could be considered my constructive thesis. This constructive thesis does not involve causal explanations that appeal to origins in answering "why" questions. Rather, this is a logical description that operates with a principle of essential necessity in its attempt to answer "how" questions concerning biological phenomena.

3.2 Kant's indispensability thesis

Kant provides an original, systematic, and influential account of organisms in the *Critique of Judgment*. One of Kant's main conclusions in the Third Critique is that judgments concerning organisms are possible through a reflective (as opposed to determinate) judgment that operates with a certain kind of necessity – the necessity involved in purposive judgments that appeal to an end or goal. He did not think that it was possible to give a mechanistic account of organisms in that mechanistic necessity does not capture what is distinctive about the kinds of organization proper to animate objects. While the Newtonian laws of motion have a mechanistic necessity that are universally lawful for inanimate nature, Kant was pessimistic about the prospects of mechanistic explanations in the life sciences. He famously states,

> For it is quite certain that in terms of merely mechanical principles of nature we cannot even adequately become familiar with, much less, explain, organized beings and how they are internally possible. So certain is this that we may boldly state that it is absurd for human beings even to attempt it, or to hope that perhaps someday another Newton might arise who would explain to us, in terms of natural laws unordered by any intention, how even a mere blade of grass is produced.[3]

Kant's point is not that there will never be progress in the life sciences, which is to say, he did not deny that mechanistic explanations of animate matter could and will never progress, e.g., due to the inherent limitations of the cognizing subject or the nature of animate matter.[4] Kant's point, rather, is that whether or not mechanical forms of explanation progress in the natural sciences, purposive reasoning concerning organisms would remain indispensable in a systematic approach to the life sciences. Generally speaking, this could be considered Kant's indispensability thesis.

3.3 Kant's antinomy of judgment

Kant's indispensability thesis emerges through his attempt to resolve the antinomy of judgment. Kant presents the antinomy of judgment in Sections 69–70 the Third Critique through the formulation of two maxims or principles of theoretical investigation that are in an apparent conflict. An antimony is a debate in which each side of dialectical conflict can successfully raise objections to the other side, but cannot successfully defend itself from the objections posed to its position. According to the first mechanical principle, "All production of material things is possible in terms of merely mechanical laws." According to the second teleological principle, "Some production of material things is not possible in terms of merely mechanical laws."[5] These maxims are formulated as constitutive principles for determinate judgment of material objects and, as such, the necessity involved in the conflict arises in reason, not the power of judgment. If these maxims were formulated in terms of the possibility of judgments, the maxims would not directly conflict with one another. The formulation proper to the faculty of judgment would concern the way in which the production of material things and their forms must be judged to be possible in terms of mechanical laws. As a conflict of the faculty of judgment, the tension between the two maxims concerning the possibility of material things in terms of mechanical laws does not obtain a necessity, which is to say, it is merely subjective and regulative. The direct conflict that generates the antinomy, by contrast, arises "within the legislation of reason."[6] This means that the two conflicting maxims "have their basis in the nature of our cognitive powers" and the logical dialectic that is generated obtains its necessity according to the logic of the categories of the understanding. In other words, the conflict regarding the possibility of mechanistic explanation does not find its source in the power of judgment, but is legislated to judgment by the faculty of reason. This means that the necessity of the conflict between the two maxims is a logical necessity, not merely a syntactical or grammatical conflict that arises in judgment.

Kant illustrates the apparent logical conflict between these two maxims through an example of bird morphology and flight.

> Moreover, so far is objective purposiveness, as a principle for the possibility of things of nature, from being connected *necessarily* with the concept of nature that it is rather this very purposiveness to which we primarily appeal in order to prove that it (nature) and its form are contingent. For when we point, for example to the structure of birds regarding how their bones are hollow, how their winds are positioned to produce motion and their tails to permit steering, and so on, we are saying that all of this is utterly contingent if we go by the mere *nexus effectivus* in nature and do not yet resort to a special kind of causality, viz., the causality of purposes (*the nexus finalis*); in other words, we are saying that nature, considered as mere mechanism, could have structured itself differently in a thousand ways without hitting on precisely the unity in terms of a principle of purposes, and so we cannot hope to find *a priori* the slightest basis for the unity unless we seek it beyond the concept of nature rather than in it.[7]

Mechanistic explanations of inanimate matter operate with a causal necessity that has a lawful universality. The laws of motion in the Newtonian sense are universally and necessarily binding on all material things. The lawful regularity of mechanistic explanations of inanimate matter are made possible by the principle of efficient causality in which a preceding event necessarily brings about an antecedent event. The principle of efficient causality grounds its logical necessity in the temporal supposition of forward succession. These features of mechanistic explanations seem to logically conflict with purposive explanations of bird morphology and flight. Judgments such as "The purpose of the bird's wings is flight" or "The lightness of the bird's hollow bones is purposive for flight" or "The purpose of the bird's tail feathers is steering" have a different logical structure than the determinate judgments of mechanistic explanations. First, explanations of the bird's morphological parts (e.g., bones, feathers, wings, and tail) are not supposed as aggregated independent parts. Rather, purposive judgments concerning the bird's flight reason in accord with the suppositions that conceive of these morphological parts as organized and interdependent. Second, the bird's morphological parts are judged in a "means-end" or "for the sake of" in relation to the large-scale organization of the bird's flight activity as a contingent whole. The two maxims thus conflict with regard to the kind of causality involved in the explanation of the bird's flight (efficient and purposive). Third, the apparent conflict between the two maxims concerning the production of material things involves different principles of necessity. Whereas mechanistic explanations of inanimate matter ground efficient causation in the necessity of forward temporal succession, judgments concerning the bird's flight involve a contingent relation that reasons from the bird's flight (end) to its morphological parts (means). Kant's resolution of the antinomy involved the formulation of a principle of necessity that can ground the necessity-amidst-contingency proper to purposive explanations.

3.4 Mechanistic explanations of material objects

Kant's argument that organisms are "natural purposes" is developed by distinguishing purposive explanations with at least three features of mechanistic explanation: 1) a material object is known through its aggregated parts, 2) efficient causality governs the relations among material parts, and 3) matter is spatially determined with lawful regularity. First, he identifies mechanistic explanations of unified objects through a reference to the independent character of their parts.[8] The objects of mechanistic explanations are wholes that are generated by independent parts joined together as aggregates. A mechanistic explanation has explanatory purchase in determinate judgments concerning aggregate collections and provides a methodological principle for natural scientific investigation – the natural scientist seeks the explanation of the unified wholes of material objects through the identification of its independent parts and the clarification of their attributes. Kant clarifies this reductive feature of mechanistic explanation through an analogy with a machine,

> For the effect we see in these machines is caused by their part not insofar as each part on its own contains a separate basis, but only insofar as all of them together contain a joint basis making these machines possible. But it is quite contrary to the nature of physical-mechanical causes that the whole should be the cause that makes possible the causality of the parts; rather, here the parts must be given first in order for us to grasp from them the possibility of a whole.[9]

This reductive feature of material composition is contrasted with the holistic features of purposive explanations in which the "whole precedes the possibility of the parts."[10] He similarly states, "if we seek the cause merely in matter, as an aggregate of many substances extrinsic to one another, then we have no principle whatever [to account] for the unity in the intrinsically purposive form of its structure."[11] In short, mechanistic explanations explain material objects in terms of their aggregated parts.

Second, Kant uses the terminology of mechanism interchangeably with efficient causal explanation. This is the sense of mechanism that is used to introduce the critique of teleological judgment in the second part of the *Critique of Judgment* where Kant states, "we are saying that all of this is utterly contingent if we go by the mere *nexus effectivus* in nature and do not yet resort to a special kind of causality, viz., the causality of purposes (the *nexus finalis*)."[12] The operative distinction here is between efficient and final causality and Kant interchanges the term "mechanism" with *nexus effectivus* (efficient causality) in mutual opposition to purposive causality. In contrast with the explanation of material objects through the independence of constituent parts, mechanistic explanations explain successive changes in a manifold, e.g., explaining an event through a reference to a necessarily preceding event. A mechanism in this sense has the mark of explanatory necessity that unifies successive changes in a manifold according to the principle of efficient causality.

Third, Kant identifies mechanistic explanations with the lawful regularity of matter. Material objects are governed by universal laws that comprise the content of the science of mechanical physics. Kant speaks of "mechanistic laws" that are the "true physical grounds of explanation" and states that "we can and should endeavor to [proceed] in terms of nature's merely mechanical laws as far as we can."[13] Mechanical explanations have a lawful regularity through which the empirical concept of matter bears the marks of necessity and universality, e.g., Newton's laws of motion. One of Kant's assumptions in characterizing the lawfulness of mechanistic explanations concerns the spatial location of material objects – they obtain a lawfulness according to their extension in space. Kant does not develop this feature of mechanistic explanation in detail in the *Critique of Judgment*, but his broader account of the empirical concept of matter is relevant in connecting the two previous features of mechanistic explanation (a material object is known through its aggregated parts and efficient causality governs the relations among material parts).

Kant develops his broader account of the concept of matter in the *Metaphysical Foundations of Natural Science* where he characterizes the science of mechanics as one of the four divisions of the system of matter. This fourfold division correlates to the headings of the table of categories: phoronomy (the science of the quantity of motion of matter) correlates with quantity, dynamics (the science of the qualities of matter) correlates with quality of the motion of matter, i.e., attractive and repulsive force, phenomenology (the science of the perceptual appearance of the motion of matter) correlates with modality, and mechanics (the science of the relation of the parts of matter to each other) correlates with relation. Kant's account of the empirical concept of matter proceeds in three stages that conform to his overall division of the study of matter. These stages of Kant's analysis correspond to three characterizations of matter in terms of motion: 1) "the movable in space," 2) "the movable insofar as it fills a space," and 3) "the movable insofar as it...has moving force."[14] Consider the second dynamic conception of matter further. Material objects and their parts are defined dynamically through a reference to spatial location – they fill a space and have external relations with other material objects with which the objects cohere and repel. Material objects have the relation of "being outside of" or "external to one another" in that they are essentially conditioned by space as the form of external relations.[15] More to the point, each material object can be divided into indefinitely smaller bits of matter, each of which fills its own homogeneous spatial location. (This spatial location is itself indefinitely divisible according to its own repulsive and cohesive forces.) Thus, not only do material objects have an externally related spatial position in relation to other material objects, but the parts of material objects are also externally related according to spatial location. In other words, material parts obtain their independence from each other in virtue of the divisibility of their essential spatial determination. It is in this sense (spatial divisibility and homogeneity) that Kant claims that matter lacks the unity of composition and is thus a merely aggregated collection, which is to say, a material object is merely comprised of an aggregate of independent parts.

This point is significant for several reasons. It connects the first and third feature of mechanical explanations involved in Kant's argument (that material objects are known through their aggregated parts and spatially determined with lawfulness). It also highlights that Kant's account of the empirical concept of matter fundamentally conceives of matter as that which fills space. Kant's conception of matter in terms of spatial location underlies the first maxim of mechanical explanation in the antinomy of judgment concerning the possibility of mechanistic explanation – that all production of material things is possible in terms of mechanical laws. The universality of this maxim is derived from the essential feature of material objects – spatial location. Not only does spatial location provide the basis for distinguishing among material objects, but it conditions the kind of unity proper to matter – the aggregation of independent parts that are, in principle, indefinitely divisible according to the homogeneity of spatial location.

In sum, mechanistic explanations of material objects have at least three features that are operative in the first maxim of the antinomy – "All production of material things is possible in terms of merely mechanical laws." We have seen that mechanistic explanations include a reference to material objects known through aggregated parts, that efficient causality governs the relations among material parts, and that the relations among material parts is spatially determined with lawful regularity. As we will see in the next section, these features of mechanistic explanations are each relevant in the contrast that Kant makes with the second maxim of the antinomy.

3.5 Purposive explanations of animate material objects

The second maxim involved in the antinomy of judgment concerns the possibility of mechanistic explanations of animate material objects. Generally speaking, Kant did not think that mechanistic explanations fully grasp what is distinctive about organisms (animate material objects). Organisms have a stronger kind of unity than the aggregation of independent parts proper to inanimate matter. For example, a branch of a tree is unified with the other parts of the tree, e.g., roots, trunk, leaves, in a different and stronger way than a bit of granite is part of a rock or an individual gear is part of a watch. The piece of granite remains a piece of granite even when spatially separated from the rock and the gear of the watch can be separated from the watch without eventually losing its shape as a metal object. But when the tree branch is cut off and spatially separated from the rest of the tree, its characteristic attributes undergo change because these attributes are connected to a stronger relationship with the tree as a whole. In short, the branch is now a branch in name only (homonymously). It becomes wood. What is distinctive about the animate matter of organisms that is not fully captured in the explanations of mechanical principles? Kant's answer to this question can be reconstructed through three essential features of organisms: 1) interdependent parts and wholes, 2) compositional unity that is contingent rather than necessary, and 3) the reciprocal means-end or purposive relations among organic parts and wholes. Let's consider these features in more detail.

First, organic parts have a dependence on organic wholes and these parts alter their properties and effects, depending on the state of the whole or the activities of the other parts. The distinctive feature of organic parts concerns their dependency with other organic parts and the whole in which they are comprised. Kant stated,

> the inner possibility of a whole as a purpose always presupposes that there is an idea of this whole and presupposes that what these parts are like and how they operate depend on that idea, which is just how we have to present an organized body.[16]

Put simply, organic parts have characteristic activities that are internally dependent on the activities of other organic parts and the continued activity of the organic whole. The parts of the bird's digestive system, for example, include organs such as a stomach, liver, kidneys, and intestines that do not operate independently, serially, or in parallel to one another, but are heterogeneously coordinated with each other and cooperate in the overall activity of the digestion of nutrients. This coordination of organic parts and cooperation in the activity of the whole has a distinctive unity that is not proper to the spatial contiguity and cohesion of the independent parts of inanimate matter. Mechanistic laws do not capture what is distinctive about organisms in that they underdetermine the particular heterogeneity or diversity of organic parts.

Second, organisms are also distinctive, according to Kant, in that their unity is contingent rather than necessary. As a tree grows, for example, its parts are produced through the activity of the whole, e.g., water is distributed from the roots to the leaves, and the leaves produce nutrients that are distributed to the further growth of the roots. The digestion of the bird, for another example, is conditioned by the activity of the bird as a whole as it forages for food. The activities of the organic parts in these examples are not only dependent in their coordinated activity, but also dependent on the activity of the organism as a whole. This double sense of dependence of organic parts introduces a contingency for further activity.[17] If the tree does not get enough water from its roots, the leaves will wither and die. The continued activity of the leaves is thus contingent on the continued activity of the roots. If the bird does not fly around and forage for food, it will die and the organs associated with its digestive activity will cease their particular activity. Organic parts are thus contingently related to organic wholes for their continued activity. The contingency of organic parts and wholes, moreover, also works the other way – the activity of organic wholes is contingently dependent on the activity of organic parts. If the tree roots do not absorb and distribute water to the other parts of the tree, the tree as a whole will cease its growth. If the bird's stomach does not absorb nutrients from the food, the bird as a whole will cease its foraging activities – it will die. In short, organisms are distinctive not only through a stronger unity of interdependence, but a unity that is contingent for the further characteristic activity. It is precisely in light of the stronger interdependence and contingency of organic parts that they are organized rather than merely aggregated.

Third, mechanical laws do not capture what is distinctive about how organisms are causes and effects of themselves.[18] Organisms are natural purposes in that the interdependent and contingent relations among parts and wholes are reciprocal means-end relations. Kant stated,

> A causal connection, as our mere understanding thinks it, is one that always constitutes a descending series (of causes and effects): the things that are the effects, and that hence presuppose others as their causes, cannot themselves in turn be causes of these others. This kind of causal connection is called that of efficient cause (*nexus effectivus*). But we can also conceive of a causal connection in terms of a concept of reason (the concept of purposes). Such a connection, considered as a series, would carry with it dependence both as it ascends and as it descends: here we could call a thing the effect of something and still be entitled to call it, as the series ascends the cause of that something as well.[19]

The relations among interdependent organic parts and wholes are means-end or purposive relations in at least two senses – ascending and descending. First, the activities of organic parts are a means to the end of the activity of the organic whole. The ascending direction of purposiveness is from part to whole, which is to say, not only are the coordinated and cooperative activities of organic parts, e.g., the digestive systems in birds or the nutrient distribution systems of trees, are related as means through which the activity of the organic whole, e.g., the flight of the bird or the growth of the tree, is possible and finds its fulfillment. The functioning of an organic part is a means to the functioning of the other parts and the whole. Second, the activity of the organic whole is the means to the end of the activities of the organic parts and it thus has a descending purposiveness. The flight of the bird or the growth of the tree (activities of organic wholes) are the means for the continuing activities of the parts in their interdependent relations. Both ascending and descending means-end relations comprise the organism's self-organization. As Kant stated, the organism's "parts [are] combined in a whole by being reciprocally the cause and effect of their form" and thus "everything is an end and reciprocally a means as well."[20] He continued, "Only if a product meets that condition [reciprocal means-end relations], and only because of this, will it be both an organized and self-organizing being...."[21] In short, the continued activity of the whole (ultimately its survival) comprises this combined activity of the parts and serves as a means for their continued activity and thus has a "self-preserving purposiveness."[22]

3.6 Kant's resolution of the antinomy

We are now in a position to return to the antinomy of judgment and the apparent conflict between the two maxims concerning the mechanistic laws of (in)animate matter. Recall that an antinomy is a debate in which each side can successfully raise objections to the other side, but cannot successfully defend itself from the objections posed to its position. Kant's attempt to resolve the antinomy identifies the source of the conflict and demonstrates that both sides of the debate have a shared false assumption. The false assumption in the antinomy concerns a univocal conception of judgment that operates in the same way for both maxims. In short, Kant resolves the antinomy by making a distinction between two kinds of judgment – determinate and reflective. In the process, he attempted to clarify the kind of logical necessity involved in how we come to know organic forms.

Generally speaking, the faculty of judgment plays a unique role in Kant's broader critical project. Along with understanding and reason, judgment is one of the higher faculties of knowledge and mediates between them.[23] But the faculty of judgment also encompasses them both and is operative in the two other faculties. The three critiques of Kant's critical project are concerned with the faculty of judgment in its theoretical features (*Critique of Pure Reason*), the judgment involved in the practical features of reason (*Critique of Practical Reason*), and aesthetic-teleological features (*Critique of Judgment)*. The faculty of judgment "in general" is the faculty of thinking the particular as contained under the universal. That is, it establishes a relation between the particular and the universal (the particular being a case or instance, and the universal being the concept, rule, or principle). Each mode of judging has a particular distinguishing feature: a theoretical judgment "contains an is or an is not" and a practical judgment contains an "ought, the necessity of why something happens for some purpose or other."[24] The task of the *Critique of Judgment* is dedicated to the clarification of purposiveness in judgments of aesthetic taste concerning pleasure and pain and teleological judgments concerning nature. Kant clarifies these judgments by attempting to establish an *a priori* principle that governs them – the principle of purposiveness.

Recall the antinomy's thesis ("All production of material things is possible in terms of merely mechanical laws") and antithesis ("Some production of material things is not possible in terms of merely mechanical laws"). The first thesis maintains that the production of material things is possible through explanatory mechanical principles, that is, mechanistic explanations provide knowledge of material things in a way that has the marks of necessity and universality. Kant states that one "ought always to reflect on these events and the forms in terms of the principle as far as one can, because unless we presuppose it in our investigation [of nature] we can have no cognition of nature at all in the proper sense of the term."[25] In short, mechanistic explanations are necessarily involved (and ought to be involved in any systematic epistemology) in the ability of reason to know nature.

However, judgments concerning organisms that have a natural purpose involve a descriptive shift of attitude. This theoretical attitude shift is motivated, in part, by one of the features of mechanistic explanations highlighted earlier – a material object is known through its aggregated parts. A mechanistic explanation that seeks to know organisms in terms of its parts, however, is limited in at least three ways. First, it is limited by its inability to grasp the interdependency involved in the self-organization of organic parts and wholes. Second, organic parts and wholes have reciprocal means-end relations that are spontaneous with each other, rather than mere successive temporally sequences of causes and effects. Efficient causality connects two successive events in time and the spontaneity of reciprocal means-end relations among organic parts and wholes confuses the necessity of this temporal sequencing. Third, the unity involved in the self-organization of organisms is contingent rather than necessary. The activities of the organic parts are not only dependent in their coordinated activity, but also dependent on the activity of the organism as a whole. This double sense of the dependence of organic parts introduces a contingency for further activity.

These features of organic form as natural purposes conflict with the features of mechanistic explanations that suppose a reference to material objects that are known through aggregated parts, efficient causality that governs the relations among material parts, and that the relations among material parts as spatially determined with lawful regularity. It is the apparent conflict among these features of mechanistic explanations and the natural purpose of organic forms, then, that one is motivated to shift to an interest in natural purpose in an attempt to find another principle that can complement mechanistic explanations. Kant states, "in the case of certain forms in nature it [judgment] has to think of their possibility as based on a principle that differs from that of natural mechanism."[26] In short, the conflict involved in the mechanistic explanations of organic form motivates a search for a principle that is compatible with mechanistic explanations, but can explain the distinctive features of natural purposes.

Kant resolves the antinomy by making a distinction between determinate and reflective judgments. Kant stated in the first introduction to *Critique of Judgment*, "Judgment in general is the ability to think the particular as contained under the universal."[27] Determinate judgments are possible through the ability to determine an underlying concept (universal) by means of a given empirical presentation (particular).[28] A determinate judgment starts from the universal concept and descends to the particular through a *specification* of the diverse under a given concept. In other words, it makes the universal concept specific. These kinds of judgments are involved in the first maxim of the antinomy ("All production of material things is possible in terms of merely mechanical laws"). By contrast, a reflective judgment is possible through thinking the particular instance in its diversity as contained under the universal concept. A reflective judgment proceeds from the empirical particular and ascends to the universal and thereby attempts to classify the diversity of the particular. Kant maintained that reflective judgment is the logical use of judgment, which is to say, it belongs to the subject's power to reflect. The principle of purposiveness belongs to reflective (not determinate) judgment in which the particular is presented and the power of judgment searches, as it were, for a universal that can apply to it. The principle of purposiveness is in this sense a heuristic that guides or regulates cognition in the mode of an "as if" but is not itself determinate.

How does Kant's distinction between the intentional stances of determinative and reflective judgments contribute to the resolution of the antinomy of judgment? We have seen that the conflict of the antinomy motivates a search for an additional principle that can capture what is distinctive about organic forms. Kant articulates this additional principle as the *a priori* principle of purposiveness involved in teleological judgments. In order to make judgments concerning what is distinctive about organisms as self-organizing, the principle of purposiveness is indispensable. This additional principle is not necessarily in conflict with mechanical explanations in that it is merely reflective, not determinate. While the principle of natural purposiveness may be indispensable in judgments concerning organic forms, it does not explain their activity.[29] Rather, reflective judgment – in its search for a universal in order to classify the diversity of the particular – posits a natural purpose as an idea that characterizes the self-organization of organic forms. This heuristic or regulative principle posits purposiveness as a regulative idea but in a process that underdetermines the diversity and particularity of material objects. The two principles can complement one another insofar as reflective judgment posits a natural purpose proper to unity of organic form and backfills, as it were, with mechanical explanations of material components of organism. Thus, there is not a direct conflict between mechanistic and non-mechanistic explanations.

Reflective judgment thus does not employ an objective principle, but a merely subjective principle or a maxim that has a logical necessity proper to the subjective conditions for thinking in general. This merely subjective necessity lacks the necessity of an objective principle due to the peculiar discursive character of human cognition. What is the peculiar discursive nature of our intellects that is involved in the attitudinal shift between mechanistic and purposive explanations in Kant's account of organic form? I submit that one possible answer to this question concerns Kant's doctrine of the schematism. Discursive reasoning applies concepts to determinate material objects of sensibility through schemata, the determination of time as necessary forward succession, and the *a priori* form of intuitions. The principle of purposiveness, however, cannot be schematized and therefore cannot become an objective principle that can be determinately predicated of material objects. Rather, it becomes relegated to a merely subjective principle that can only be reflectively applied to objects.

I have argued that Kant's account of organic form plays a decisive role in how Kant frames the antinomy of judgment and have reconstructed three essential features of this account, namely, that organisms have 1) interdependent parts and wholes, 2) compositional unity that is contingent rather than necessary, and 3) organic parts and wholes that can be properly characterized in reflective judgments as reciprocal means-end or purposive relations. We have seen, in particular, that the unity conditions concerning self-organized material objects (organisms) play an important role in distinguishing between inanimate and animate matter – mechanistic explanations do not capture what is distinctive about the compositional organization of organic wholes and parts.[30] How are organisms unified in such a way that resists the kinds of explanations according to aggregated parts? Kant does not provide a full answer to this question in the *Critique of Judgment*, but this fuller answer is provided by his account of the way in which the "synthetic unity of the understanding" unifies sensible manifolds in the *Critique of Pure Reason*. Once we have clarified the formalism in Kant's account unity of material objects in general – without regard to their aggregate or compositional structure – we will be positioned to clarify Kant's epistemological justification for the unity that is specific to organic forms.

3.7 Kant's account of unity in *Critique of Pure Reason*

There are two predominant senses of unity in the *Critique of Pure Reason* – a critical sense as the first of the categories of quantities and a traditional transcendental sense. In his discussion of the table of categories in the Transcendental Analytic, Kant recognizes these two senses of unity in his discussion of the twelve categories and their relation to the scholastic transcendental notions of being, unity, truth, and the good. These notions are transcendental in the sense that they transcend Aristotle's categories and must be accounted for extra-categorically, which is to say, they are notions that signify that all things participate in or partake of these transcendental determinations.[31] While the transcendentals were traditionally regarded through extra-categorical determinations, Kant initially incorporates them into the categories of quantities. Kant states, "These supposedly transcendental predicates of things are, in fact, nothing but logical requirements and criteria of all knowledge of things in general, namely, unity, plurality, and totality."[32] He distinguishes a categorical and transcendental sense of unity, however, when he states,

> This unity, which precedes *a priori* all concepts of combination, is not the category of unity, for all categories are grounded in logical functions of judgment, and in these functions combination, and therefore unity of given concepts, is already thought. We must therefore look yet higher for this unity, namely in that which itself contains the ground of the unity of diverse concepts in judgments, and therefore of the possibility of the understanding, even as regards it logical employment.[33]

While Kant generally sought to restrict the analytic notion of unity to the category of quantities, he nevertheless encountered problems with this restriction. One problem with this restriction arose in the context of Kant's attempt to derive the table of categories where he encountered a need for a unifying activity of the understanding.

This cannot be the unity of the category of quantity because it is precisely the categories that Kant is attempting to derive in the first place.[34] The category of unity already presupposes a broader ability to cognize the synthetic accomplishments involved in judging. Thus, Kant must resort to providing a different, extra-categorical (transcendental) account of the synthetic unity of judgments.

Kant's transcendental account of the unity of synthetic manifolds can be introduced through the parallel between the sensible manifold and its unity with the separation of the faculties of understanding and sensibility. The faculty of sensibility provides an intuition of sensible manifolds that, in themselves, are not unified. A purely sensible intuition here is a raw and chaotic stream of sense data. As Kant stated, "But the combination of a manifold in general can never come to us through the senses, and cannot, therefore, be already contained in the pure form of sensible intuition."[35] Rather, we come to have knowledge of sensible manifolds through the combinations and analysis that are the result of an act of the understanding. As Kant stated, "all combination…is an act of the understanding."[36] A sensible manifold is synthesized by the ability of the understanding to combine elements in the manifold. Indeed, Kant even characterizes the understanding as "nothing more than the power to combine *a priori* and to bring the manifold of given intuitions under the unity of apperception."[37] The understanding (cognizing subject) accomplishes the synthesis in a given manifold and thereby makes possible and confers the unity of intuitions. This is the case in both sensible and conceptual intuitions – the unity of a sensible synthesis provides intuitions of objects and the unity of an intellectual synthesis provides concepts of objects. In both cases, to the degree that intuitions are synthesized manifolds, they have a unity that is the accomplishment of the cognizing subject. I submit that the formalism in Kant's account of the constitution of unified synthetic manifolds concerns this one-sided dependence on the understanding of the cognizing subject. This is the precise sense in which Kant's account of organic forms supposes an epistemological idealism.

Kant's transcendental account of the unity of synthesized manifolds ultimately appealed to the self-consciousness of the cognizing subject – the unity of apperception. He made this appeal in order to ground the synthesizing accomplishments of the knower in a necessary condition that makes possible unified synthetic manifolds. The proposition "My representations are mine" is an analytic statement that nevertheless reveals a synthesized manifold. I am empirically aware of myself as I make representations. I am the same subject throughout a variation of representations. The "I" has a "mineness." My identity as the same subject is a necessary relation throughout the manifold of various representations.[38] As Kant stated, "I am conscious of the self as identical in respect of the manifold of representations that are given to me in an intuition, because I call them one and all my representations, and so apprehend them as constituting one intuition."[39] Kant called this self-consciousness of one's identity in the process of representation the "unity of apperception" – I apperceive my own identity as a cognizing subject as I contribute to the process of representation. Kant stated, "in the transcendental synthesis of the manifold of representations in general, and therefore in the synthetic original unity of apperception, I am conscious of myself, not as I appear to myself, nor as I am in myself, but only that I am."[40] This is not an empirical self-awareness, but a self-consciousness that has a necessity – insofar as I am involved in the constitution of objects, I apperceive the unity of my own identity (this is different than representing various aspects of myself in reflective self-awareness). The "I think" (*cogito*) accompanies my representations and thereby contributes an original synthetic unity to the process of cognizing synthetic manifolds. The cognizing knower legislates that manifolds are synthesized such that their unity is coherently thought by the subject through the means of concepts. In short, the constitutive origin of the unity of synthesized manifolds is the transcendental unity of apperception – it is the unifying condition that makes knowledge of sensible and conceptual manifolds possible.

Kant's transcendental account of the unity of synthesized manifolds reasoned from the "condition" to the "conditioned." How does the unity of apperception – the "I think" that necessarily accompanies all my representations – condition the synthesis of manifolds? We have seen that Kant did not think that sensible manifolds have a unity that originates in sensibility. He maintained a strict separation in his faculty psychology between the understanding and sensibility. Sensible manifolds are fundamentally contingent and Kant's argument in the Transcendental Deduction concerning the constitutive priority of the unity of apperception does not imply that the cognizing subject confers unity, as it were, onto completely arbitrary sensible intuitions at will. Sensible intuitions are receptive, but they are not merely receptive in that they have the forms of intuition of space and time. Sensible manifolds, in other words, are spatio-temporal manifolds that conform to the outer (spatial) and inner (temporal) forms of sensible intuitions. The intuitions of space and time, for Kant, are forms that limit the way in which the given is manifest in appearances. But form does not necessarily imply unity for Kant. Sensible intuitions of spatio-temporal manifolds are not, in themselves, unified manifolds. Kant states,

> Thus unity of the synthesis of the manifold, without or within us, and consequently also a combination to which everything that is to be represented as determined in space or in time must conform, is given *a priori* as the condition of the synthesis of all apprehension – not indeed in, but with these intuitions. This synthetic unity can be no other than the unity of the combination of the manifold of a given intuition in general in an original consciousness, in accordance with categories, in so far as the combination is applied to our sensible intuition.[41]

In the application of the categories to a sensible intuition, the unity of apperception makes possible the syntheses of spatial manifolds. Kant's example is the perception of a house – I perceive the house in a spatial manifold that is necessarily unified. The house is a material object that is located in a spatial manifold. Upon reflection, I can abstract the form of space as a homogeneous system of univocal locations in a formal intuition. The intuition of spatial manifolds is synthesized through the application of the *a priori* concept of homogeneity, one of the concepts that comprise the category of quantity proper to the understanding. Homogeneity of spatial location has a logically necessary unity that is obtained by 1) the application of the category of quantity and 2) the unity of apperception that conditions the synthesis of categories. I apprehend the sensible object (house) in a spatial manifold as filling a determinate location in a system of homogeneous locations that "must necessarily be in conformity with the synthesis of apperception."[42] In short, the unity proper to synthesized spatial manifolds of the intuition of space obtains its necessity through the application of the categories and the synthetic unity of apperception. The unity of synthetic spatial manifolds is given "not indeed in, but with these intuitions." That is, the representation of spatial manifolds is accompanied by the unity of apperception that, along with the categories, synthesizes the aggregate of spatial location as a necessarily unified spatial field that is populated by material objects.

The unity of apperception also makes possible the synthesis of temporal manifolds. Kant's example is the appearance of freezing water in which two states – fluidity and solidity – have a temporal relation to each other. The "before" of the water's fluidity and "after" of the water's solidity are relations in a temporal succession, which is to say, the temporal manifold is synthesized as forward succession. Upon reflection, we can abstract the form of time from the perception in a formal intuition that reveals the role of the unity of apperception in the determination of the temporal intuition as necessary forward succession. The "I think" that accompanies the temporal manifolds contributes a necessary connection between "before" and "after" – my representations of fluid water were mine previously and my representations of the solid are now also mine. The "mineness" of the unity of apperception provides the identification through which the temporal manifold is synthesized as a successive "before" and "after." Kant stated, "I necessarily represent to myself synthetic *unity* of the manifold, without which that relation of time could not be given in an intuition as being *determined* in respect of a time-sequence."[43] The unity of apperception provides the continuity through which a temporal manifold is synthesized as forward succession. Time, the inner form of sensible intuition, is a unified synthetic manifold that conditions the application of the category of efficient causation to the freezing water. The freezing temperature (cause) and the solid water (effect) are related to each other in a successive manifold of "before" and "after." This application of the category of causality to the sensible intuition of the freezing water illustrates how the unity of apperception accomplishes the synthesis of temporal manifolds. The "I think" that accompanies all my temporal representations provides the continuity through which temporal phases are synthesized.

We have seen that Kant's account of the unity of synthesized manifolds is constitutively grounded in the unity of apperception and application of the categories to sensible (spatio-temporal) manifolds. He reasons in the Transcendental Deduction from the unifying condition ("I think") to the conditioned sensible manifolds through an account of how the sensible intuitions of space and time are necessarily unified as a homogeneous system of univocal locations (intuition of space) and forward succession (intuition of time). Kant has not, however, completed his account of the unity of synthesized manifolds. To complete his account, Kant needs to answer the question of how the pure concepts of the understanding (categories) are applicable to intuitions of sensible objects in a way that relates to sensible objects in their determinate particularity. Kant's examples of the perception of the house and freezing water illustrate how the unity of apperception and the categories are applicable to the form of sensible intuitions (spatio-temporal manifolds), but the examples do not illustrate how we have knowledge of objects in their particularity, e.g., the real actuality of the house and freezing water as material objects.[44] The objective validity of the knowledge of determinate objects in a sensible intuition involves a matching of the intuition and a concept. Intuitions without concepts are blind and concepts without intuitions are empty. Kant's account of the unity of synthesized manifolds is not complete until he addresses how sensible intuitions match concepts, e.g., how the sensible intuition of a house or freezing water matches the concept of house or freezing water. He addresses this issue through his theory of the schemata.[45] As I argue later, the schemata play an important role in Kant's distinction between determinate and reflective judgment in *Critique of Judgment*. Indeed, Kant employs his theory of schemata to distinguish between these two kinds of judgments and thereby resolves the antinomy of judgment. Judgments concerning natural purposes are merely reflective, he argues, because natural purposes cannot be schematized in a determinate judgment.

3.8 Schemata in the *Critique of Pure Reason*

A determinate judgment concerning sensible manifolds matches an intuition and a concept through schemata, which Kant introduces in the Analytic of Principles as "third things" that mediate between intuitions and concepts.[46] The pure concepts of the understanding (categories) enter into and contribute to the concepts that correlate with intuitions of particular objects. To the degree that they do contribute to the intelligibility of representations, is the degree in which our knowledge can be characterized with necessity. This process of "entering into" and "contributing to" the match with intuitions is incomplete in that there is a mismatch between the categories as intellectual principles devoid of images, and the sensible image proper to a sensible intuition. The third thing (schema) that mediates between the categorized concept and intuition must be sufficiently homogeneous with both, yet devoid of the defining characteristics of both (spontaneity and receptivity). A schema functions in two directions – it prepares the intuition for the determination of the concept (*subsumption*) and adapts the concept for application to intuition (*application*).[47] First, it prepares the intuition for the determination of the concept through the work of the imagination. Strictly speaking, raw intuitions of sensible manifolds are chaotically received and the active part of the faculty of sensibility – the imagination – prepares the sensible intuition for conceptualization by organizing the sense data into intelligible appearances in an apprehension. This involves the combination of impressions that are sensibly received in an intuition into more or less assembled and coherent appearances. Kant is clear that the process of schematization works through the imagination in the direct perception of appearances, e.g., the house with its specific shape, color, and magnitude. This function of preparing sensible intuition for conceptualization underlies and makes possible the employment of the imagination in a reproduction of a sensible object as an image, e.g., a recollection of one's childhood home through the work of the imagination in memory. In short, the process of schematization involves preparing the sensible intuition for conceptualization through the identification of the particular attributes of the object, e.g., the house's shape, color, and size, and searching for a universal concept for which they can be subsumed. Second, the process of schematization functions to adapt the concept for application to an intuition. This second function involves the application of the categories as principles for concept formation and revision. Schemata in this sense function as rules that guide and, indeed legislate, the application of concepts to intuitions. Kant illustrates this function of the process of schematization through an example of the concept of "dog," "The concept 'dog' signifies a rule according to which my imagination can delineate the figure of a four-footed animal in a general manner, without limitation to any single determinate figure."[48] The schematization of concepts of particular intuitions such as the concept of a dog involves the identification and clarification of the conceptual features of dogs in general (four-footed canine mammal) and these features are principles that guide the conceptualization of the appearance of a dog in a sensible intuition. This involves a process of adaptation in the trial and error at work in assimilation, e.g., I may come across a three-legged dog and have to qualify the universality of my generalizations. In the process, I am attempting to make the concept of dog specific to the particular instances that I might possibly encounter. In short, this second function of schemata involves the application and adaptation of concepts to intuitions by reasoning from the universal principle proper to the concept of the particular instance.

Not only do schemata function in the concept formation of particular intuitions, e.g., a house or dog, but they also play a more important role in Kant's transcendental account of the unity of synthesized manifolds in that they realize the synthesis of the categories (accomplished by the unity of apperception) in the imagination's apprehension of particular intuitions. Kant characterizes a schema as a "rule of synthesis of the imagination" according to which the understanding provides the categories as principles or rules that guide and should govern the process of conceptualization universally. In this transcendental sense, the schemata are the application of the pure concepts of the understanding (categories) as principles for the imagination. Kant lists the "rule like" application of the categories roughly as follows: the schemata of the category of quantity is number – a unity of the synthesis of a quantitative manifold. The categories of quality are being in time (reality), not being in time (negation), and the same time both filled and empty (limitation). The schemata of relation are "permanence of the real in time" (substance), "the real upon which, whenever posited, something else always follows" (causality), "the co-existence, according to a universal rule, of the determinations of the one substance with those of the other" (community or reciprocity).[49] And the schemata of modality are "the determination of the representation of a thing at some time or other" (possibility), "existence in some determinate time" (actuality), and "existence of an object at all times" (necessity).[50] Generally speaking, these schemata express the categories as principles and are themselves organized as they are synthesized with each other in the process of cognition. They each realize the category in a representation as a "determination of the representation of a thing at some time or other," which is to say, they are all involved in the temporal combination of events in forward succession. Kant states, "In the synthesis of appearances the manifold of representation is always successive."[51] The schema of causality, for example, is a rule that objectively determines the temporal relation between events in a forward succession, e.g., freezing water that changes from fluid to solid. That one event always follows from another event in forward temporal succession is an application of the category of causality to sensible objects universally. The "always" here is the mark of the schema of necessity that combines with the temporal determinations of the causal relation.

Temporal determinations are also involved in the schema of substance – permanence of the real in time. This schema refers to the lasting connection between the permanent substantial bearer of qualities. The permanence, like the "always" of the schema of causality, bears the mark of necessity that characterizes the connection between a thing and its attributes across a merely momentary temporal connection in intuition. The temporal connection between a substance and its qualities persists beyond the momentary intuition. This temporal determination of the schema of substance can be clarified by an example of a concept of a particular sensible object, e.g., the perception of a house. When I make the judgment – The house (substance) is brown (quality) – I am synthesizing a substance and its quality in a way that persists beyond the immediate perception. It is in this sense of permanence that the schema of substance is temporally determined.

Temporal determinations are fundamental in the process of the schematization of the other categories as well. Generally speaking, the process of applying the categories (pure concepts of the understanding) to sensible intuitions involves temporalizing them to conform to the inner form of intuition – time (necessary forward succession). Kant states,

> The schemata are thus nothing but *a priori* determinations of time in accordance with rules. The rules relate in the order of the categories to the time-series, the time-content, the time-order, and lastly to the scope of time in respect of all objects.[52]

This schematization of the categories through the temporal determinations of intuitions involves a "transcendental synthesis of imagination" that is "simply the unity of all the manifold of intuition in inner sense [time as the inner form of intuition], and so indirectly the unity of apperception which as a function corresponds to the receptivity of inner sense."[53] What kind of unified syntheses are involved here? It is clear from earlier in the section that the schemata contribute to the unity of synthesized manifolds through the application of the categories with temporal determination. The categories are themselves synthesized by the unity of apperception in the understanding. But the unity of apperception reappears in Kant's discussion of the work of the imagination in bringing the concept and intuition together in a temporal determination. What kind of synthesis does the "I think" (*cogito*) contribute here? Kant is not clear. I submit that the unity of apperception functions here in a way that corresponds to the necessity of forward succession. I am implicitly self-conscious "that I am" through a series of successive temporal determinations of intuitions. The continuity across successive temporal phases of a representation is ensured by the unity of apperception – I can know that the change in liquid to freezing water is connected because I apperceive myself in the perception of the liquid (before) and solid water (after) – the "I think" temporally manifests in the perception of the two states and remains as a retentional phase within them such that a temporal manifold can be synthesized across momentary phases. Kant stated,

> As the grounds of an *a priori* necessary unity that has its source in the necessary combination of all consciousness in one original apperception, they [schemata] serve only to subordinate appearances to universal rules of synthesis, and thus to fix them for thoroughgoing connection in one experience.[54]

Time is a form of intuition that applies to any intuition whatsoever, even to the inner intuition that I have of myself ("that I am") in the unity of apperception. The categories are realized in an intuition through the temporalizing process of schematization, a temporal process that itself is a synthetic manifold that is unified by continuity afforded by the unity of apperception.

3.9 Schemata in the *Critique of Judgment*

We are now in a position to return to Kant's distinction between determinate and reflective judgments in the *Critique of Judgment* and to investigate the role that the schemata play in this distinction. We have seen earlier that Kant relegates judgments concerning natural purposiveness to reflective judgments rather than determinate judgments. Recall that his argument involved an account of mechanistic explanation and organic form. As highlighted earlier, the mechanistic explanations of determinate judgments involve at least three features of explanation: 1) a material object is known through its aggregated parts, 2) efficient causality governs the relations among material parts, and 3) matter is spatially determined with lawful regularity. Kant's general argument is that mechanistic explanations do not explain what is distinctive about organisms. We have also seen that organisms are distinctive in at least three senses according to Kant's account of organic form: 1) interdependent parts and wholes, 2) compositional unity that is contingent rather than necessary, and 3) organic parts and wholes can be characterized in reciprocal means-end relations. Mechanistic explanations do not fully explain what is distinctive about organisms – namely, the reciprocal relations of the organism's contingent compositional unity (heterogeneous parts that are interdependent with each other and their wholes). In order to capture what is distinctive about organisms, it is necessary to appeal to the principle of purposiveness in reflective judgments.

This argument still remains incomplete. To complete the argument, Kant needs to further demonstrate how the form of organisms is not adequate to the mechanistic explanations of determinate judgments – particularly with regard to the kind of unity involved in heterogeneous parts as functionally diverse. Kant completes his argument by appealing to his theory of schematism – which he characterizes in the *Critique of Judgment* as the peculiar (*Eigentümlichkeit*) character of human cognition. The discursive character of cognition involves the synthesis of intuitive manifolds through the categories and unity of apperception through the application of concepts to intuitions through schemata. The principle of purposiveness, however, cannot be schematized and, as a result, cannot be applied to the sensible intuitions of determinate judgments. As we will see later in the section, the principle of purposiveness cannot be schematized primarily due to 1) the temporal determination (necessary forward succession) that is fundamental to the process of schematism that becomes confused by the reciprocal means-end relations among organic parts and wholes, and 2) the absence of specificity (underdetermination) regarding the unity of heterogeneous parts as diverse.

Kant develops the role of schematism in the distinction between determinate and reflective judgments in the first introduction to the *Critique of Judgment* in a section entitled "On Reflective Judgment." Kant's basic argument is that while determinate judgments schematize concepts and intuitions in a way that provides mechanistic explanations of inanimate matter, reflective judgments cannot schematize the principle of purposiveness in a way that provides the same kind of necessity involved in mechanistic explanations.[55] Kant begins this argument by making a distinction between determinate judgment that has the "ability to determine an underlying concept by means of a given empirical presentation" and reflective judgments that have the ability to "reflect on a given presentation so as to [make] a concept possible."[56] Determinate judgments start from the universal concept and descend to the particular through a specification of the diverse under a given concept. Note that Kant's characterization of determinate judgments corresponds to the second function of schema as rules that guide and legislate the application of concepts to intuitions. Knowledge of inanimate nature is possible in that its appearance corresponds to the cognitive achievements of the knower. Determinate judgments schematize the categories and unity of apperception in the application of a concept to a sensible intuition. They are not in need of a special principle (such as the principle of purposiveness that is needed by reflective judgment) because they are in conformity with the principles of understanding that make possible the unity of sensible manifolds. Kant states,

> understanding and judgment schematize *a priori* and apply these schemata to each empirical synthesis, [the synthesis] without which no empirical judgment whatever would be possible. Here judgment not only reflects but also determines, and its transcendental schematism also provides it with a rule under which it subsumes given empirical intuitions.[57]

Not only do determinate judgments apply concepts to sensible intuitions through the process of schematization, but the schematization also has the function of preparing the intuition for conceptualization through reasoning from the particular to the universal. That is, the schematization of concepts and intuitions in determinate judgments also discursively involves the first function of schematization that proceeds from the empirical particular and ascends to the universal and thereby attempts to classify the diversity of the particular.[58] Both functions of schematization are operative in determinative judgments, "judgment not only reflects but also determines...." But in determinate judgment, e.g., "This house is brown," we are predicating the concept "brown" to the particular house and "classifying" the presentation of the house through the identification of its conceptual color determination. Determinate judgments thus subsume the particular in an underlying concept.

The schematization involved in determinate judgments temporalizes the pure concepts of the understanding (categories) and presents principles through which sensible intuitions become intelligible. We have seen, for example, that the schematized principles of causality (the real upon which one event follows from another) and substance (permanence of the real in time) involve forward succession as a temporal determination. The temporalized principles subsume sensible intuitions as spatio-temporal substances with causal relations and so on. Kant's argument is that the spatio-temporal materiality of sensible intuitions is confirmed, in theoretical cognition, by the "universal concepts of nature"[59] in which the lawful regularity of the principles is universally applied. That is, upon further reflection on nature in general, we can recognize that the principles have universal applicability. Newton demonstrated that nature can be known with a universal lawfulness that confirms the lawfulness of particular sensible intuitions, e.g., the house as a spatio-temporal substance with causal relations and so on. Determinate judgments can thus have an objective validity – confirmed in its universality – that inanimate matter is lawful in its appearance. Mechanistic explanations are determinate judgments that employ the necessity and universality of the phenomenal world of classic mechanics.

The process of schematization involved in judgments concerning sensible intuitions of animate matter is different. Organic forms of animate matter, as we have seen, have 1) diverse parts that are interdependent with each other and the wholes they comprise, 2) compositional unity that is contingent rather than necessary, and 3) parts and wholes that can be characterized in reciprocal means-end relations. In the attempt to make a determinate judgment regarding animate matter with organic forms, the process of schematizing a concept with a sensible intuition is limited. The activity of the whole is an end for the activity of the parts at the same time as the activity of the whole is a means for the further activity of the parts. Moreover, the activity of the parts is the means for the activity of the whole at the same time that the activity of the parts is the end for the further activity of the whole. For example, the organs of a dog's digestive system, e.g., intestines, kidney, and liver, are the means of the dog's behavioral activity, e.g., running, jumping, and eating, while simultaneously being the end of the behavioral activity, e.g., eating as means of digestion. The reciprocal means-end relations of organisms resist the schematization of determinate judgments in that these relations do not temporally appear in a forward (irreversible) temporal succession, but co-exist in a simultaneity that can only be distinguished through a shift to a reflective attitude.[60] Reciprocal means-end relations thus confuse the schematization that temporalizes the categories as principles that are fundamentally characterized through necessary forward succession. Reciprocal means-end relations, by contrast, have "co-existence" as their temporal modality and judgments that are temporally schematized in the temporal modality of forward succession do not obtain a determinate grip on the sensible intuition of the functioning organism. The temporal modality of co-existence that characterizes the reciprocal means-end relations of organisms cannot be objective in a world that is fundamentally known through the temporal modality of forward (irreversible) succession. Through this difference in temporal modality involved in the process of schematization, Kant begins to demonstrate his distinction between determinate and reflective judgments. The temporal co-existence of dynamic reciprocal means-ends relations of organic parts and wholes cannot be schematized in a determinate judgment that temporalizes concepts fundamentally according to forward succession.

The process of schematization involved in judgments concerning sensible intuitions of animate matter is also different with regard to the heterogeneity or diversity involved in the organic forms of animate matter. Determinate judgments regarding organisms underdetermine the heterogeneity or diversity involved in the interdependence of organic parts and wholes. Kant attributes the absence of specificity (underdetermination) regarding the unity of heterogeneous parts as diverse to the "parsimony" or frugality that is "suitable for our judgment."[61] This frugality of judgment with regard to the specification of interdependent relations among diverse parts and their whole underdetermines this diversity. The activity of the whole organism is intimately connected to the activity of the diverse parts, but it is difficult to see how these diverse parts – in their purposive heterogeneity – can be understood without reference to the activity of the whole. Kant illustrates this underdetermination of individual organisms as complex systems with heterogeneous parts through an analogy with a taxonomic approach to animate nature in its totality. When we survey the "exceedingly heterogeneous" forms of organisms in nature, it is practically impossible to conceptualize the commonality among organisms while maintaining a specificity regarding the vast array of the heterogeneity of organic forms. Kant asked,

> How could we hope that comparing perceptions would allow us to arrive at empirical concepts of what different natural forms have in common, if nature, because of the great variety in its empirical laws, had made these forms (as is surely conceivable) exceedingly heterogeneous, so heterogeneous that comparing [them], so as to discover among them an accordance and a hierarchy of species and genera, would be completely – or almost completely – futile?[62]

Even if it were possible to provide a universal taxonomy of the diversity of organic forms, the classifications would need some kind of principle as a guide.

Kant here takes a jab at Carl Linnaeus in a footnote regarding his difficulty of finding a principled genus and species for his taxonomic system.[63] This principle would need to be something different from the schematization of the categories but compatible with it. It would be a logical principle for a reflective judgment that generalizes the diversity of interdependent relations among parts and wholes in functional descriptions.

The principle of purposiveness is a "subjective principle" that is conditioned by practical reason's hypothetical and retrospective cognition. In the purposeful action of an agent with practical reason, the concept of a future state of affairs can temporally precede and bring about a change in the present state of affairs. For example, I conceptualize the money I would earn if I built and rented out a house and then I build the house and collect the rent money.[64] Intentional agents practically reason in light of ends and this teleological activity, for Kant, is an "ideal causality" that provides a principle for reflective judgment. Reflective judgments appropriate the purposiveness of practical reasoning as their principle from which to retrospectively infer (and underdetermine) the means-end relations of diverse organic parts and wholes. The principle of purposiveness is different from the schematized categories, however, in that it can merely be stated as a necessary presupposition. It cannot assert its principle as a law – it is not "nomothetic" like determinate judgments can be. The principle of reflective judgments is merely a principle for the "logical use of thought" in that purposive relations among organic parts and wholes cannot be schematized – they do not have an irreversible temporal modality of forward succession and they underdetermine the specificity of heterogeneous organic parts.

Both reflective and determinate judgments generally consist of bringing a concept and intuition together in a synthesis. In a schematic exhibition of a determinate judgment, a concept is applied to a corresponding intuition such that the unity of concept and the intuitive manifold are matched in an *a priori* lawfulness that is confirmed by the lawfulness of nature in the Newtonian sense. Kant contrasts this schematic expression of determinate judgments with the symbolic expressions of reflective judgments in which "there is a concept which only reason can think and to which no sensible intuition can be adequate, and this concept is supplied with an intuition that judgment treats in a way merely analogous to the procedure it follows in schematizing."[65] Whereas schemata contain direct exhibitions of the concept that are demonstrative, symbols contain direct exhibitions of the concept that are analogous. Reflective judgments can symbolically and indirectly exhibit the principle of purposiveness by proceeding according to two logical directions (similar to double functioning of schemata in synthesizing concepts and intuitions). Symbols can proceed from 1) a universal sense of purposiveness to the specification of particular natural purposes and regard "nature as art" or 2) a "technic of nature" that renders natural purposes as a result of the intentions of a designer (God). It can proceed empirically and ascend from the particular to the universal and thereby classify the diverse through a comparison of common characteristics.[66] This process of classification in reflective judgment involves comparisons among particulars into empirical concepts (classes) and generalizes more common laws (higher genera) so as to arrive at an empirical system of nature. This would perhaps look a lot like Linnaeus' taxonomic system, but with a logical principle of purposiveness that organized classifications according to functional descriptions rather than the morphological differences among botanical forms as sensible shape. Such a logical system of organisms, however, would nevertheless remain an indirect and symbolic exhibition with regard to intuitions of particular organisms. It would remain "artificial" in that it would operate with a mere "subjective principle" rather than an objective one.[67] Reflective judgments that classify organisms proceed from particulars in search of purposive conceptualizations – the concept of natural purpose functions as a regulative heuristic that can guide, be confirmed, or disconfirmed, by subsequent determinate judgments. We have seen earlier that the principle of purposiveness in reflective judgments that logically classify organic forms are grounded in the subject's ability to reflect on reasoning in accordance with an end and retrospectively generalize this "in order to" or "means-end" relation to the self-organized relations of organic parts and wholes.

3.10 Clarifications: two levels of conceptualization

We are now in a position to step back from Kant's argument and attempt to generally characterize his account of organic forms and how they are known. Kant's account operates with a logically sophisticated view of the kinds of unity proper to organisms. At a basic epistemological level, organisms are given in sensible intuitions in much the same way as inanimate objects. They are spatio-temporal material objects that can be partially known with the same kind of necessity as inanimate objects. Consider again Kant's example of a dog in the *Critique of Pure Reason* – the dog appears in a sensible intuition as a manifold of sense data that is synthesized through a process of schematization such that the concept "dog" and sensible particular are brought together in a judgment. The dog can be known through the laws of Newtonian science that provide mechanistic explanations that are conditioned and made possible by the structures of cognitions (i.e., pure concepts of the understanding, the intuitions of space and time, and the unity of apperception). In the process of schematization, the appearance of the dog in the sensible intuition and the empirical concept of a dog as a four-legged canine mammal are brought together in a synthesized manifold that is unified in a one-sided constitution rooted in subjective accomplishments. We have here two levels of conceptualization that can be easily confused – the empirical concept of dog and the pure concepts of the understanding. This point is worth clarifying in that Kant's account of organic form in the *Critique of Judgment* is primarily concerned with the identification and clarification of an *a priori* principle proper to the faculty of judgment (the principle of purposiveness). As we have seen earlier in the section, the role of schematization in Kant's argument for the distinction between determinate and reflective judgment is best understood in light of his account of the unity of synthesized manifolds in the *Critique of Pure Reason*. Kant's broader account of the unity of organisms implies – and must imply – that there are two levels of conceptualization that are operative in cognition generally and the cognition of organic forms in particular. This implies that we have empirical concepts of dogs, cats, maple and beech trees that can be schematized with sensible intuitions in a way that condition the mechanistic explanations that are consistent with determinate judgment. Determinate judgments concerning organisms operate with schematized empirical and pure concepts that condition and make possible mechanistic explanations of living phenomenon. The role of schematization in the distinction between determinate and reflective judgments is oriented by Kant's concern with comprehensiveness – his indispensability thesis maintains that any complete account of organic forms must include purposive judgments that find their *a priori* necessity in the principle of purposiveness. Recall Kant's argument concerning the total diversity of organisms – any conceptualization of organic forms in their totality necessarily over-determines the heterogeneity of organic forms. The conceptualization of living things in their totality (as animate matter in contrast with inanimate matter) is exhausted by the overwhelming diversity of organic forms. But an *a priori* principle that has a legitimate mark of necessity would have universal purchase that is proper to the totality of living things. The diversity of organic forms over-determines the efforts of determinate judgments to reach such universality. We reach a conceptualization of organic forms, rather, through the adjustment of the intentional stance between determinate and reflective judgments. We can conceptualize the forms of animate matter in their totality through reflective judgments that logically account for the distinctive features of organic forms: 1) interdependent parts and wholes, 2) compositional unity that is contingent rather than necessary, and 3) organic parts and wholes as reciprocal means-end or purposive relations that can be properly characterized in reflective judgments.

Kant's argument in the *Critique of Judgment* concerning the limits of the universality of organic diversity in determinate judgments does not sufficiently incorporate the role of empirical concepts that his broader epistemology entails. The argument serves to ground reflective judgments in an *a priori* principle that is proper to the faculty of judgment (just as the understanding and practical reason have their own principles). Through this prioritization of the establishment of *a priori* necessity, Kant's arguments minimize the legitimate role of empirical concepts that his overall account entails. Empirical concepts are operative in both determinate and reflective judgments. As we have seen, determinate judgments schematize empirical concepts and sensible intuitions in a way that conforms to inner intuitions (temporality) and outer intuitions (spatiality), e.g., dogs are spatio-temporal material objects that are subject to the same laws of Newtonian mechanics as inanimate objects. Empirical concepts are also involved (and must be involved, given Kant's broader account) in purposive descriptions, e.g., the purpose of the dog's heart is blood circulation and the purpose of the dog's kidney is waste extraction. Here, the empirical concepts of blood circulation and waste extraction characterize the immanent purposiveness of the dog's organs (heart and kidney) in light of the activities of the whole – blood circulation contributes to the process of oxygenation and respiration and waste extraction from the blood contributes to the digestion of nutrients. Blood circulation and water filtration are empirical concepts of the activities of the parts that contribute to the activities of the whole (respiratory and nutritional systems). Empirical concepts of organic forms can thus be both mechanistic and purposive. This point clarifies Kant's indispensability thesis – he did not maintain that we do not have empirical concepts of organic forms that can organize systems of classification. Kant recognized the possibility and legitimacy of reasoning from particular immanent purposes in a system of classification proper to reflective judgments. Such a system of classification inevitably utilizes purposive reasoning of empirical concepts. Rather, Kant's primary concern in his account of organic form in the *Critique of Judgment* involved a search for a kind of necessity proper to such a system of classification of organic forms. This is the sense in which he dismisses botanical taxonomy as lacking a principle that has necessity and universality. As we have seen previously, Kant's arguments for the *a priori* principle of immanent purposiveness attempts to provide a necessity that is grounded in the synthetic achievements of the cognizing subject. The unity of organic forms – of heterogeneous parts that have a reciprocal relation to the activity of organic forms – is not fully conditioned by the unidirectional temporality involved in forward succession. Rather, the necessity of reflective judgments concerning immanent purposiveness is conditioned and made possible by the logical synthesis of the pure concepts of the understanding and the unity of apperception.

3.11 Unified manifolds in general: synthesis or identity?

So far, this investigation has attempted to present a sympathetic reconstruction of Kant's basic account of organic forms and how we come to know them as purposively and immanently self-organized. I have attempted to distill his account to its epistemological foundations and have relegated Kant's broader metaphysical commitments. Like Husserl, I have an epistemological reading of Kant rather than a metaphysical one. My focus has been on the way that Kant accounts for the organized unity of sensible manifolds proper to organic forms. My critique of Kant and initial development of a phenomenological notion of biological form draws upon resources from Husserl's phenomenology. By way of transition, allow me to briefly and generally contrast Husserl's phenomenology with Kant's critical project.

Husserl's relationship to Kant was complex and developed throughout his career. Following Franz Brentano, Husserl's early work was initially dismissive of Kant, but after 1905 he began to engage Kant's critical project in his lectures and seminars. Husserl was particularly attracted to Kant's transcendental idealism and he agreed with Kant that naïve objectivism ignores the role of the achievements of the cognizing subject. Husserl's phenomenology and Kant's critical project both attempt to account for objectivity in terms of the conditions of possibility for objectivity and both projects are committed to the view that the cognizing subject is involved in the disclosure of objects in the world precisely as experienced by empirical subjects in the world.[68] Husserl credits Kant as seeing the problematics of transcendental philosophy clearly – more clearly than his predecessors, e.g., René Descartes. But through his engagement in debates with the Neo-Kantian schools of his day, Husserl also sharply distinguished his phenomenological project from Kant's critical philosophy, developed criticisms of Kant's epistemology, and rejected distinctions that are inherent in Kant's transcendental idealism, e.g,, Kant's faculty psychology (sensibility, understanding, and reason), his distinctions between *a priori* and *a posteriori* knowledge, analytic and synthetic judgments, and appearances with things in themselves.[69] While Husserl did not systematically engage Kant's account of organic form, his criticisms of Kant's positions regarding the synthesis of manifolds and the intentional stance of the observer provide important implications for a phenomenological approach to organic form.

An initial point of entry into the differences between a Husserlian-inspired approach to the unity of organic forms and Kant's account can perhaps be most directly and clearly seen through contrasting their accounts of the unity of synthesized manifolds involved in perception. Husserl radicalized Kant's insight that a synthesis of a manifold is that which "first gives rise to a cognition" (CPR A77/ B103). However, Husserl's disagreements with Kant with regard to the unity of synthesized manifolds ultimately involved two distinguishable epistemological positions that have a cascade of implications for a phenomenological account of organic forms.

Kant's supposition that the intelligible sources of the unity of sensible syntheses does not lie in the objects themselves does not merely lack the justification that his critical project demands. It also conflicts with phenomenological evidence. The unity of sensible syntheses is not reducible to the cognitive achievements of the knower, that is, they are not ultimately grounded in a psychological process that is prior to and independent of discrete, non-intentional sense impressions. Rather, perception involves the direct grasp of unified manifolds (complex wholes) that, by contrast, are independent of higher order cognitive achievements such as judgments or representations that mediate the "real actuality" of the unity of perceptual objects. Consider again Kant's example of the visual perception of a house. The house is given in perception as a unified manifold, e.g., with a front and backside that are gradually presented as one walks around the house. The house is given through a series of profiles through which the house obtains its objectivity. How is it that this manifold of visual profiles (appearances) presents one and the same object?

One answer to this question might involve appealing to the qualitative matching of the house profiles, e.g., the color, size, and shape of the house. Each visual profile presents qualitative features of the house in slightly modified profiles that are synthesized as an individual object with a unity that is based upon the associative similarity of the visual qualities. Qualitative matching across the manifold is a necessary condition of the house as a unified manifold, but according to Husserl's account of the unity of manifolds, it is not a sufficient condition.

An additional and stronger kind of unity is accomplished and here we see the strong epistemological realism of Husserl's account of the unified manifolds. It is not merely that the perceptual manifold has an associative synthesis of like-with-like, but that the house presents itself as an identity-in-a-manifold, that is, as an identical object with its own intrinsic unity proper to itself. The house is perceived as one and the same identity given amidst the modifications involved in the manifold of visual profiles generated by walking around the house. This unification of identity is proper to the object and not accomplished in an act of synthesis of the cognizing subject. Technically speaking, the unity of identification is not even a synthesis insofar as a synthesis brings together individuals that are determinately distinguishable. The unity that is presented in perceptual objects is an identity – one and the same individual – in each of the manifold of appearances. The appearance does not mediate between the thing in itself and the cognizing subject, but rather the thing in itself in its real actuality is pre-given in the appearance. However, this is not to say that subjective accomplishments are not involved in the process of unification. The embodied perceiver conditions the constitution of perceptual objects by generating the manifold. As one walks around the house, for example, the modifications of visual profiles generate a series of visual profiles through the movements involved in embodied perception. The embodied "I," in other words, generates the visual manifold through its own movement. But the work of identification is accomplished by the thing itself and is genuinely discovered (not produced). The thing is disclosed with its own essential and unified sense thoroughly and from the beginning. We thus have an account of the unity of synthesized manifold that is two-sided (in contrast to Kant's one-sided account).

Husserl characterized Kant's one-sided account of unified manifolds as "based on essential misunderstandings" that misconstrued synthesis as the "exclusive accomplishment of the understanding."[70] As Husserl stated, "Kant failed to notice that many combinations of content are given to us where no trace of a synthesizing activity that produces connectedness is to be found."[71] Kant's supposition that the intelligible origin of the unity of manifolds does not lie in sensible intuitions but is rather an active accomplishment of the understanding prevented him from seeing what Husserl took to be obvious and significant – that "experience, and it alone is decisive here, shows nothing of such 'creative' processes."[72]

Husserl thus distanced his phenomenological application of a theory of manifolds from Kant's as early as the *Philosophy of Arithmetic* and remains cautious with regard to the characterization of cognitive processes in terms of "mental acts" or "activities." He maintains this caution in the *Logical Investigations* where he stated, "In talking of acts, we must steer clear of the word's original meaning: all thought of activity must be rigidly excluded."[73] The characterization of the unity of perceptual objects in terms of psychological processes or mental acts betrays a basic presumption of the natural attitude. The sense in which perceptual objects disclose themselves as identities-in-a-manifold – as unified in their particularity as individuals – is not explainable merely in terms of the synthesizing accomplishments of the cognizing subject but is thoroughly and from the beginning rooted in the unified sense in which organisms are disclosed as identities-in-a-manifold.

3.12 Self-organizing manifolds: a logic of sense

This difference between Kant and Husserl's accounts the unity of manifolds of perceptual objects gains philosophical significance in application to the complex organization of organic forms. Living and non-living objects are given in perception in fundamentally the same way – as identities-in-manifolds with an intrinsic unity of individuality. These unified manifolds do not merely present the bare individuality of objects, but they present complex objects with features that are organized according to various founding-founded relations of presentational dependency. The initial perceptual difference between animate and inanimate objects concerns an additional layer of sense in which animate objects are disclosed with self-organized activities that make sense of their environment. Generally speaking, animate things have a more intimately integrated or immanently organized individuality than inanimate things, an individuality that varies according to a sense-making process in its environment. A living thing such as a cell, fungi, plant, or animal is given an identity-in-a-manifold that has an added layer of self-organizing and habitual activity. At this general and descriptive level of analysis, many of the traditional accounts of the kind of changes involved in growth and development of living things have a logical validity, e.g., living things have an internal principle of change, self-generation, and the like. More formally, living things are not merely presented through the unity of an identity-in-a-manifold (with an individual "thisness"), but as a complex object with an organized and organizing unity that is disclosed through its habitual and adaptive activity in its environment.

In thematizing the self-organization and habituation of organic forms as an added layer of sense proper to individuals, I am not considering an object that is ontologically distinguishable from the perceptual attributes of its spatio-temporal materiality, but the same identity-in-a-manifold in its organization as a complex object. More specifically, perceptual objects in general present identities-in-a-manifold with organizational features that contribute to the heterogeneity of an object's unity and can be analyzed in terms of founding-founded relations. This is the case for both inanimate and animate perceptual objects. Both kinds of objects have a perceptual sense that is not merely the unity of identification, but the unity of association that is proper to its perceptual givenness, e.g., color cannot be presented without brightness, brightness cannot be presented without surface, surface cannot be presented without shape, and so forth. The perceptual sense of objects of experience is unified with a necessity proper to individuality (identities-in-a-manifold) and an additional layer of necessity that is proper to the constitution of its organization. I develop Husserl's conception of the kinds of necessity that are proper to the organization of complex objects in the next investigation, but I submit at this point that the presentational senses of complex perceptual objects in general are given through a logic of sense in which essential necessities are disclosed. The essential necessities proper to the logic of sense of perceptual objects are also operative in the presentational sense proper to the functional organization of organisms.

The logic of sense that is operative in the perceptual givenness of perceived objects is also operative in the self-organization involved in habitual sense-making in an environmental situation. In directing one's theoretical interest to the distinctive features of living things, the distinctively biological attributes of organisms are presented with organizations that have unities that are given in passive syntheses with a logic proper to themselves. This logic of sense of self-organized and habitual variations of organic forms does not derive its necessity from the conditions of possibility of the cognizing knower, but the organic forms "in their own right" and have a necessity that is not coincidental to the sense in which they are disclosed as what they are. The notion of form here is broader than that found in Kant's account – it does not merely include an organism as an atomistic individual externally related to an environment, but an organism with sense-making activities that are internally unified (and thereby essentially conditioned and constrained) by variations in its environment. Organisms have intimately unified relations of self-organization and these relations are internally unified with (not merely coupled in an external relation to) the sense-making processes involved in the habitual and adaptive activities in their environment. This phenomenological articulation of organic form is different from Kant's account in that it operates with a theory of intentionality that addresses the problem of necessity through an account of objective sense rather than appealing to the synthesizing achievements of the cognizing knower. In other words, this phenomenological move shifts the issues involved in Kant's "grounding problem" and addresses these issues through a thicker notion of the organized sense of biological individuals. For example, the necessities involved in the formal claim – organisms are always already "organisms-in-an-environment" – can be grounded through a process of reasoning from particular and diverse biological individuals to various taxonomic generalizations such that the claim obtains a universality proper to all life forms. The necessary unity here is proper to the objective sense of organisms in the biological world and its justification can be tested by appealing to particular empirical examples. Questions concerning generalizations in biology can thus be empirically answered and the logic of sense proper to organisms accordingly have morphological essences rather than exact ones. In short, this object-oriented approach to organic forms grounds the problems of logical necessity in the objective sense in which an organism is disclosed as an identity-in-a-manifold with a functional sense that can be morphologically characterized through a logic of habitat fitness.

The distinctive methodological features of a phenomenological approach to the question of organic forms become further apparent in the application of part-whole logic to the organizational features of organisms. What kinds of part-whole relationships are involved in the self-organization of biological individuals? This question can be answered in different ways, depending on the suppositions of the intentional stance of the investigator. Generally speaking, there are at least three preliminary positions regarding the ontological suppositions of part-whole relations in organic self-organization. First, we can explain the functional organization of an organism through its parts. The activities of an organism as a whole can be explained through accounts of the activities of the organism's parts. This bottom-up approach starts with the supposition that biological parts are ontologically basic to biological wholes. These suppositions of part-whole relations could be defined as follows: the whole is nothing but the combination of the parts in that the whole is intelligible through the associations among parts and does not have intrinsic existence without them. An organism is a so-called "bottom-up whole" in the various respects that correspond to various reductionistic approaches:

1) The material of whole is nothing but the material of its parts.
2) The state or property of the whole is nothing but the state or property of the parts and their combination.
3) The functions of wholes are nothing but the functions of parts and their combinations.
4) The efficient causes of wholes are nothing but the efficient causes of the parts and their combination.[74]

There are various kinds of reductionism in biology that can be distinguished according to the ontological suppositions involved in these kinds of part-whole relationships. These suppositions condition how functional questions get answered, namely, through an appeal of the function of lower level or basal conditions of biological parts, e.g., organism, organ, tissue, cell, chromosome, DNA, gene, protein, and so on. Each lower level of analysis provides the answers to the functional questions at the higher level. These "bottom-up" approaches are also operative in multi-level explanations of ecosystem ecology as it was formulated, for example, by Eugene Odum. According to Odum, ecological sciences consider part-whole relations that are proper to the individual, population, community, ecosystem levels. "Bottom-up" approaches suppose that the holistic attributes of ecological communities can be explained in terms of the attributes of its populations which, in turn, can be explained in terms of the attributes of its individuals. In short, ecological units of explanation are the "sum" of their parts.

Second, we can explain the functional attributes of an organism through appeals to the self-organization of the organism as a whole. The activities of an organism's parts can be explained through accounts of the activities of the whole. This "top-down" approach is involved in, for example, questions concerning the two-sided causal relations among different levels of organization. This holistic approach asks questions along with Roger Sperry concerning the emergent properties of the bird's flight.[75] When the bird takes flight, it brings its constituent parts with it. The large-scale organization operative at the level of the organism has a complexity that does not seem to be compressible to the lower levels. These incompressible causal networks, while causally dependent on the lower levels, introduce a level of complexity with a self-organization that is not reducible to the lower levels. In contrast to the "bottom-up" approach, this holistic approach considers the emergent properties of an organic whole as a distinct level of organization in virtue of its self-organization. The ontological suppositions involved in holistic approaches start with the activities of an organic whole and explain the function of its parts through appeals to the organization of this whole. In an ecological context, for example, the activities of the ecological community condition and constrain the activities of populations which, in turn, condition and constrain individual organisms. In short, ecological units of explanation are "more than the sum" of the units at the lower level of analysis.

A third option regarding the ontological suppositions regarding part-whole relations of the functional analysis of organisms can be provided by a more symmetrical model to part-whole relations involved in a phenomenological approach. A symmetrical part-whole logic like Husserl's is ontologically neutral when it comes to the real existence of parts or wholes but rather considers ontological dependency in presentational terms. This shift in theoretical attitude is animated by an interest in the correlations among parts and wholes and, in particular, to the lawful regularity of necessary associations between, for example, micro-level parts and higher-level organizations. By taking questions of ontological dependency in terms of modal presentational contents, theoretical interest is freed from the ontological suppositions involved in reductionist and holistic approaches to causal explanations and secures the privilege of examining necessary relations among biological parts and wholes as carrying the explanatory weight. This is precisely the sense in which phenomenological questions concerning organic forms are primarily "how" questions concerning structural organization rather than "why" questions concerning causal origins. In other words, when you start with a realist ontology of parts, the demands of theoretical coherency require you to give an explanation of how wholes emerge from parts. This theoretical requirement works the other way as well – if you start with a realist ontology of wholes, you eventually are required to give an explanation of how parts emerge from wholes. Moreover, the ontological suppositions involved in these causal approaches motivate "why" questions that are involved in the search for origins. These ontological suppositions regarding part-whole relationships help clarify the explanatory features of different kinds of functional attributes.

3.13 Conclusion: functional analysis in proximate explanations

The ontological suppositions (belief modalities) involved in the intentional stance of the biological investigator can illustrate the differences among the kinds of functional explanation that are at work in Mayr's distinction between proximate and ultimate explanations. In particular, the suppositions involved in the three approaches to the part-whole relations of organic forms sketched earlier (bottom-up, top-down, and phenomenological) illustrate the different conceptions of function involved in aetiological explanations.

Mayr's 1961 article "Cause and Effect in Biology" used an extended example of bird migration that illustrates different kinds of biological explanations. He developed his account of biological explanations in response to two different but related questions concerning bird migration.

> Let me illustrate the difficulties of the concept of causality in biology by an example. Let us ask: What is the cause of bird migration? Or more specifically: Why did the warbler on my summer place in New Hampshire start his southward migration on the night of the 25th of August?[76]

Warbler species belong to the passeriformes order of contemporary ornithological taxonomies. This order is distinguished from other aviary orders according to features of bodily morphology, specifically the arrangement of their toes – passeriformes have three toes pointing forward and one backward toe. This morphological structure is particularly well suited for perching on small tree branches and other horizontal structures with narrow diameters. Passeriformes are also called "perching birds" for this reason. During the summer months, warbler species are abundant in the northeastern forests of North America where they can be found perching in trees while they hunt for insects. Warblers are among the many songbirds that travel long distances during their annual South–North migration. There are at least five warbler species that have been identified and studied in the New England forests: cape may (*Dendroica tigrina*), myrtle (*D. coronate*), black-throated green (*D. virens*), blackburnian (*D. fusca*), and bay-breasted (*D. castanea*).[77] These species of warblers eat similar insects and have similar migratory patterns. What kinds of biological explanations are available that address Mayr's "why" questions concerning bird migration, in general, and the individual warbler's migration, in particular? Mayr itemizes several kinds of explanations that could be reconstructed as follows:

1) Nutrient Provision Explanation – The variation involved in the bird's migration is proximately correlated with a decrease in affordances for nutrient provision, e.g., mature insects that provide nutritional content to the warbler. While there are enough mature insects in the summer for the warbler to successfully raise chicks, the decline in nutrient affordance in late August varies in correlation with the developmental cycle of insect populations. The variation involved in the warbler's migration, in short, is influenced by the habitat variations involved in the warbler's community interaction with insect populations. As Mayr simply puts it, "The warbler, being an insect eater, must migrate, because it would starve to death if it should try to winter in New Hampshire."[78] In short, the warbler's migration is due to fact that the warbler is an insect eater and insect-eating birds migrate according to the seasonal nutrient provision abundance of its habitats. Mayr classifies the notion of causality in this kind of explanation as an "ecological cause."[79]

2) Genetic and Evolutionary Explanation – The warbler's migration could also be explained according to the correlation between variations of the warbler's genetic information and the variations of subsequent generations of warbler populations in New England deciduous forests. Throughout its evolutionary history, the genotypic and phenotypic traits that program the warbler's response to variations in its environment were passed onto subsequent generations. The increased phenotypic variation of the individual warbler correlates with the subsequent population abundance according to the adaptive fitness of the genetic traits selected in the variation. By contrast, a screech owl that shares a tree with Mayr's warbler does not migrate – it is a permanent resident of the New Hampshire forest. In the course of the remote evolutionary history of screech owl species, the genetic variations of

phenotypic traits that condition migration responses were not adaptive and therefore not selected. This explanation of bird migration in general can also address Mayr's second "why question" concerning the ideographic individuality of Mayr's warbler in its edaphic habitat. The warbler is an insect eater and its migration is an adaptive response to seasonal scarcity conditions involved in the nutrient provisions of its seasonal habitat. Explanations made possible by the modern synthesis of molecular and evolutionary biology extend nutrient provision explanations in the explanation of comparative differences among diverse species with regard to their adaptive fitness. The warbler migrates because it is an insect eater and, by contrast, a screech owl has adapted to eat rodents and thus has sufficient nutrient provision to survive the winter conditions in New Hampshire. While genetic explanations establish a common historical origin that provides a necessity to the associative similarity of warblers and screech owls, evolution by natural selection provides a necessity in its explanation of the associative differences between the seasonal migration habits of the two species. The warbler migrates because the genes that organize its digestive system as an insectivore have been naturally selected rather than the genes that organize the digestion and eating habits of the screech owl.

3) Diurnal and Temperature Explanation – Bird migration can also be explained by the variations involved in daylight duration and temperature. Birds on average migrate in response to the decreased amount of daylight and the seasonal decline of summer temperatures. This explanation of bird migration in general is derived from the strong statistical correlation of the variations in the bird's "intrinsic" physiological readiness to migrate and the "extrinsic" daylight and temperature variations. The correlation of the bird's readiness to migrate (optimal nutrition) and macro-ecological variations such as light and temperature provide an explanatory principle to Mayr's "why" questions that he terms "intrinsic" and "extrinsic" explanations. These explanations not only address the general question of bird migration through an appeal to these "extrinsic" variations, but can also address the individuality of Mayr's warbler, that is, they can address the ideographic question that indicates the "thisness" or individuality of the warbler. In short, this specific warbler migrates as soon as the hours of daylight and temperature have dropped below a certain level. This explanation could have stronger or weaker senses of necessity. On one end of the spectrum, the explanation of the warbler's specific migration has a strong principle of necessity that is provided by behavioral ethology and conceives of migration behavior in mechanistic stimulus-response. The decrease in daylight conditions triggered a primed response in the warbler such that it actually took flight on August 25 – the day of these daylight or temperature variations. In short, this warbler always migrates on the day with a specific quantity of daylight or temperature threshold in a mechanistic stimulus-response behavior. On the other end of the spectrum, the explanation of the individual warbler could operate with a weaker notion of necessity. When the ornithologist explains the individual warbler's migration in terms of macro-ecological variations, she

is operating with statistical measurements that are strongly suggestive of the necessity involved in efficient causation, but tracks these variations according to a statistical average that does not fully account for or is not fully sensitive to a specific individual or specific threshold of the daylight or temperature variation. The warbler's migration on August 25 does not have a necessity in the mechanistic sense that it "cannot be otherwise" but a probabilistic relation to macro-ecological variations with a necessity that remains generalized and is not sensitive to individual cases. On average, the majority of warblers migrate during the last week in August through the first two weeks in September – the actual day of the migration and specific decrease in temperature or daylight is contingent. Thus, daylight and temperature variation can address both of Mayr's "why" questions concerning bird migration in general and the "why" question concerning the individual warbler specifically with stronger or weaker notions of necessity.

Taxonomies of biological explanation such as this, according to Mayer, can be divided into two basic kinds of causality. Proximate causes "govern the responses of the individual (and his organs) to immediate factors of the environment" while ultimate causes "are responsible for the evolution of the particular DNA code of information with which every individual of every species is endowed."[80] The explanations involved in diurnal and temperature variation are proximate and the causality involved in genetic and evolutionary explanations are ultimate. Genetic and evolutionary explanations are ultimate in the sense that they answer "why" questions concerning warbler migration through an appeal to the origins of the material constituents of genes and natural history. The modern synthesis between molecular and evolutionary biology provides an "ultimate" explanation of bird migration that secures a necessary relationship of the correlation of long-term common origins of genetic heredity and the primary cause of the diversity of organic forms – natural selection.[81] Genetic and evolutionary explanations of bird migration are ultimate in the sense that they answer "why" questions by providing accounts of the genetic and evolutionary origins of bird migration. The account of origins provides the grounding for the universality at work in the explanation – warbler species migrate because the DNA codes that program migratory behavior were naturally selected and passed on through the evolutionary heredity of warblers.

By contrast, the proximate explanations provided by diurnal and temperature variation do not answer Mayr's two "why" questions with an appeal to origins. Rather, proximate explanations have a more static explanation of organisms as they are directly related to their immediate environment. Proximate explanations reason from the physiological features of particular individuals and the ideographic variations involved in the individual's environment to make more generalized claims to variously scaled collectives or groups. Proximate explanations often address "why" questions through describing "when" and "how" the change occurred. Explanations of bird migration that appeal to diurnal and temperature variations target the specific external conditions of the seasonal changes in New

Hampshire during late August. These variations are the proximate conditions in the warbler's migration particularly with regard to the temporal determinations involved in asking "when" warblers migrate.

The explanations involved in nutrient provision, however, are not as clearly distinguished into "proximate" and "ultimate." Mayr claimed that the biological explanations that appeal to ecological causes are "ultimate" because warbler nutrient provision is a complex eco-systemic process with an evolutionary history that operates through the principle of natural selection. As Mayr stated, "lack of food during the winter...[is] an ultimate cause."[82] Indeed, there is an important sense in which explanations of bird migration that appeal to nutrient provision make appeals to origins. For example, the warbler species migrate South to North to take advantage of the increased insect abundance in order to ease the burden of feeding young chicks – migration is an adaptation that has evolved through the selection of genes proper to warbler species that program the adaptive behaviors involved in migration. Warblers that migrated had increased nutrient provision and were thereby more reproductively fit – they could raise more chicks if they migrated. This explanation incorporates nutrient provision in the search for answers to "why" questions through accounts of genetic and evolutionary origins, but it does not treat nutrient provision itself as an answer to the "why" questions. The nutrient provision contributes a coherency of ultimate explanations, but does not itself operate as the source of the changes involved in warbler migration.

However, there are at least two senses in which appeals to nutrient provision are proximate as well. First, the nutrient provision involved in the warbler's habitat is proximate insofar as it appeals to the immediate abundance or scarcity of insects in the warbler's immediate forest habitat on the particular day in late August. Nutrient provision explanations, in other words, are proximate in the same sense as other proximate causes – they are sensitive to the particularity of individuals, which is to say, they can be ideographic. *This* warbler in *this* deciduous forest could not survive the winter if it lacked sufficient nutritional affordances to sustain its habitat relation. Proximate explanations, in short, can be ideographically descriptive and the ultimate explanations that incorporate nutrient provision into their account progressively lose the specificity of this feature of proximate explanations.

Second, explanations of bird migration through an appeal to nutrient provision can be proximate insofar as they help explain the physiological and functional features of the birds as self-organized. Having raised her chicks, the warbler fattens up for her upcoming flight on the large hatches of late summer insects so that she will have enough energy for the long range and difficult migration. The bird's digestive system converts the insects into stored energy for the extreme conditions of the trip and when she has reached a certain threshold in her bodily mass – she migrates. Nutrient provision is a direct necessary condition for the warbler to store up enough fat for the migration. The warbler migrates, in short, because by August 25 she stored up a sufficient amount of energy for the long and difficult journey and this optimal nutritional state is necessarily conditioned by variations in insect abundance in New Hampshire during late August. Thus, explanations of bird migration that appeal to nutrient provision can also be proximate explanations insofar as they condition the physiological and functional features of warblers.

These two senses in which bird migration can be proximately explained through nutrient provision can indicate several phenomenological features of biological forms. I submit that Mayr's nutrient provision explanations of bird migration suppose a notion of biological form and that a phenomenological approach provides an attractive alternative to a Kantian approach. My dialogue with Kant in this investigation has indicated several phenomenological features of biological forms:

1) Biological forms are given as identities-in-a-manifold of change with an essential unity proper to individual organisms.
2) The forms of biological individuals are given with an organized presentational sense that has functional attributes among the organism's parts and wholes.
3) The logic of the functional sense of biological parts and wholes supposes an internal unity (rather than an external relation) between an organism and its habitat.

These essential features of my notion of biological form can be clarified through a description of the necessary relations of nutrient provisions and bird migration. How does Mayr's warbler migrate? I submit that the warbler's nutritive sense-making involves the internal unity of the warbler's physiological variation and habitat variation.

In conclusion, this investigation has covered a lot of philosophical territory. Please allow me to only briefly take stock of the general flow of this analysis and provide some comments. We have seen in the previous investigation that the modern science of plant geography as it was pioneered by Humboldt operated with a complex notion of form comprised of botanical, organic, and spatial senses. In particular, we have seen that Humboldt's logic of plant associations operated with a conception of organic form in the Kantian sense. Does Humboldt's question concerning the geographical distribution of plants necessarily presuppose a Kantian account of organic form? That is, is there something inherent to the question itself that has decided assumptions regarding the unity proper to individual organisms that, in turn, is generalized to the collective attributes of plant associations? My answer is "no" to both questions. The question concerning the geographical distribution of plants does not logically necessitate a Kantian answer. The plant geographer does not need to suppose a notion of synthesis that suffers from a one-sided formalism. The question concerning the geographical distribution of plants is not doomed from the beginning toward idealism.

The analytic radicality of this investigation was necessary for a variety of reasons. Kant is a towering figure whose long shadow is cast over the history of 19th century German life sciences and plant geography, in particular. Kant's account of organic form was a basic presupposition that framed the initial explanatory principles and classification systems of 19th century plant geography. The radicality of this analysis is partly warranted by this significant historical influence. It is also warranted by the concern to identify the specific error that opens the door toward an uncritical idealism in biological and ecological inquiry. I have argued that that error is the one-sided notion of associative synthesis involved in Kant's theory of manifolds. My immanent critique of this presupposition was also radical in that Kant's account of organic form is complex and systematic. In order to authentically clarify the basic assumptions and methodological principles in Kant's analysis of organisms in the *Critique of Judgment*, it is necessary to contextualize his account in the *Critique of Pure Reason*, specifically Kant's theory of schemata. The radicality of this investigation is partially due to the logical and systematic rigor of Kant's critical project in general. Finally, the intimacy of this dialogue with Kant has been radical in the sense that some of the analytic tools that Kant employed are reinterpreted in the initial sketch of my phenomenological alternative, e.g., part-whole logic, theory of manifolds, synthesis, and so on. The next investigation continues this reinterpretation of some of these basic analytic resources.

This investigation has yielded several important philosophical results. Kant's account of organic form puts into relief several initial features of a phenomenological approach to the logic of biological and ecological forms. This investigation clarifies the key methodological difference between a Kantian and phenomenological approach in the philosophy of biology – alternative applications of the theory of manifolds to cognitive processes and alternative accounts of synthesized associations. In the next investigation, I develop some of the basic formal features of my phenomenological approach through a retrieval of Husserl's theory of manifolds and part-whole logic. Thus far, I have simply attempted to provide a foothold for this alternative approach and traced out its explanatory implications through an engagement with Mayr's taxonomy of biological explanations from the perspective of the practicing biologist. A symmetrical or double-sided approach to the presentational sense of biological parts and wholes provides a pathway to a logic of sense of the self-organization of biological individuals. The explanatory purchase of this phenomenological alternative for the practicing biologist looks like something similar to Mayr's account of proximate explanations. That is, I agree with both Kant and Mayr that proximate explanations of biological functions are indispensable. The biological sciences have always and will always assume a logic of parts and wholes and my wager is that phenomenological resources provide an attractive alternative to Kant's approach. More substantively, my engagement with Mayr's questions concerning bird migration clarify at least three basic features of biological forms that are different from the Kantian approach. The next investigation is even more radical than this one – it develops and extends the results of this investigation and fills in the logical space that this immanent critique of Kant's account has opened up.

Notes

1 For more extensive treatments, see Hannah Ginsborg, *The Normativity of Nature: Essays on Kant's Critique of Judgement* (Oxford: Oxford University Press, 2015); Hein van den Berg, *Kant on Proper Science: Biology in the Critical Philosophy and the Opus postumum* (Dordrecht: Springer, 2014); Rachel Zuckert, *Kant on Beauty and Biology: An Interpretation of the Critique of Judgment* (Cambridge: Cambridge University Press, 2007); Jennifer Mensch, *Kant's Organicism: Epigenesis and the Development of Critical Philosophy* (Chicago: University of Chicago Press, 2013); Michael Friedman *Kant and the Exact Sciences* (Cambridge: Harvard University Press, 1998); John H. Zammito, *The Genesis of Kant's Critique of Judgment* (Chicago: University of Chicago Press, 1992).

2 There have been several recent attempts to appropriate Kant's epistemology to address the problem of functional organization in biology. Marcel Quarfood, *Transcendental Idealism and the Organism: Essays on Kant* (Stockholm: Almqvist & Wiksell, 2004); Denis M. Walsh, "Organisms as Natural Purposes: The Contemporary Evolutionary Perspective," *Studies in the History and Philosophy of Science* 37, no. 4 (2006): 771–791; Alix A. Cohen "Kant's Antinomy of Reflective Judgment: A Re-evaluation," *Teorema: Revista Internacional de Filosofia* 23, no. 1 (2004): 183–197; Angela Breitenbach, "Two Views on Nature: A Solution to Kant's Antinomy of Mechanism and Teleology," *British Journal for the History of Philosophy* 16 (2008): 351–369; Hannah Ginsborg, "Kant on Understanding Organisms as Natural Purposes," in *Kant and the Sciences*, ed. Eric Watkins (Oxford: Oxford University Press, 2001), 231–58. For recent explicit appeals to Kant's epistemology in the philosophy of biology more broadly, see Alvaro Moreno and Matteo Mossio, *Biological Autonomy: A Philosophical and Theoretical Inquiry* (Dordrecht: Springer, 2015). For a critique of the appropriation of Kant's approach in a contemporary context, see John Zammito, "Teleology Then and Now: The Question of Kant's Relevance for Contemporary Controversies Over Function in Biology," *Studies in the History and Philosophy of Science Part C* 37, no. 4 (2006): 748–770.

3 Immanuel Kant, *Critique of Judgment*, trans. Werner S. Pluhar (Indianapolis: Hackett, 1987), §75: 283. Cf. §77: 294. References to the *Critique of Judgment* will include section numbers followed by Pluhar's page numbers.

4 "Now it is entirely possible that some parts in (say) an animal body (such as skin, bone, or hair) could be grasped as accumulations governed by merely mechanical laws. Still the cause that procures the appropriate matter, that modifies and forms it in that way, and that deposits it in the pertinent locations must always be judged teleologically. Hence everything in such a body must be regarded as organized; and everything, in a certain relation to the thing itself, is also an organ in turn." Kant, *Critique of Judgment*, §66: 257.

5 Ibid., 267.

6 Ibid., 267.

7 Ibid., 236.

8 "When we consider a material whole as being, in terms of its form, a product of its parts and of their forces and powers for combining on their own…then our presentation is of a whole produced mechanically." Ibid., 408.

9 Ibid., FF IX, 236.

10 Ibid. 236; also see 408.

11 Ibid., 421.

12 Ibid., 360.

13 Ibid., FF IX, 235.

14 Immanuel Kant, *Metaphysical Foundations of Natural Science*, trans. James Ellington (New York: Library of Liberal Arts, 1970), 480, 496, and 536 respectively. Zuckert pro-

vides a helpful discussion of mechanistic explanations in *Kant on Biology and Beauty*,
101–107. My discussion is indebted to her more extensive treatment.

15 Immanuel Kant, *Kants gesammelte Schriften V, Akademie Ausgabe* (Berlin: de Gruyter,
1902), 4201; c.f. Zuckert, *Kant on Biology and Beauty*, 105.

16 Kant, *Critique of Judgment*, 408.

17 See Ibid., 360, 370, and 396, 406.

18 Ibid., 372.

19 Ibid., 372.

20 Ibid., 373 and 376 respectively. See Zuckert, *Kant on Biology and Beauty*, 123;
Zammito, "Teleology Then and Now," 218. Kant addressed these reciprocal means-
end relations in another context as well, "Now the concept of an organic being is this:
that it is a material being which is possible only through the relation of everything
contained in it to each other as end and means (and indeed every anatomist as well
as every physiologist actually starts from this concept). Therefore a basic power that
is effectuated through an organization has to be thought as a cause effective according to ends, and this in such a manner that these ends have to be presupposed for
the possibility of the effect." "On the Use of Teleological Principles in Philosophy,"
in *Anthropology, History, and Education*, ed. Günter Zöller and Robert B. Louden
(Cambridge: Cambridge University Press, 2007), 181.

21 Kant, *Critique of Judgment*, 374.

22 V: 424. See also Zuckert, *Kant on Biology and Beauty*, 119; and Zammito, "Teleology
Then and Now," 218.

23 Kant, *Critique of Judgment*, 4.

24 Kant, *Lectures on Logic*, ed. and trans. J. Michael Young (Cambridge: Cambridge
University Press, 1992), 376. See Howard Caygill, *A Kant Dictionary* (Oxford:
Blackwell, 1995), 267.

25 Kant, *Critique of Judgment*, 387.

26 Ibid., 268.

27 Ibid., 179.

28 Ibid., See also FF, 214.

29 Ibid., 360.

30 "What is the unity of the basis [that accounts] for the combination, in this product, of
the manifold [elements] extrinsic to one another? But this question, as far as it is teleological, is answered sufficiently if we posit that basis in the understanding of a producing cause that is a simple substance. If, on the other hand, we seek the cause merely in
matter, as an aggregate of many substances extrinsic to one another, then we have no
principle whatever [to account] for the unity in the intrinsically purposive form of its
structure." Kant, *Critique of Judgment*, 420.

31 Thanks to Dan Dwyer for bringing this point to my attention.

32 Immanuel Kant, *Critique of Pure Reason*, trans. Norman Kemp Smith (Boston:
Bedford/St. Martin's, 1929), 118.

33 Ibid., B131.

34 "This unity, which precedes *a priori* all concepts of combination, is not the category
of unity; for all categories are grounded in logical functions of judgment, and in
these functions combination, and therefore unity of given concepts, is already
thought." Ibid.

35 Ibid., B 130.

36 Ibid.

37 Ibid., 137.

38 Ibid., B132.

39 Ibid., B195.

40 Ibid., B157.

41 Ibid., B161.

42 Ibid., B162n.
43 Ibid., B163.
44 Kant states, "Transcendental philosophy has the peculiarity that besides the rule (or rather the universal condition of rules), which is given in the pure concept of understanding, it can also specify *a priori* the instance to which the rule is applied." Ibid., B174.
45 "The categories, therefore, without schemata, are merely functions of the understanding for concepts; and represent no object." Ibid., B187.
46 Ibid., A138/B177.
47 Ibid., A138/B177.
48 Ibid., A141/B180.
49 Ibid., A143/B183.
50 Ibid., A144/B184. See Caygill, *A Kant Dictionary*, 361.
51 Ibid., A198/B243.
52 Ibid., A145/ B185.
53 Ibid., A146/B185.
54 Ibid., A146/B186. Kant states in another context, "For the original apperception stands in relation to inner sense (the sum of all representations), and indeed *a priori* to its form, that is, to the time-order of the manifold empirical consciousness. All this manifold must, as regards its time-relations, be united in the original apperception" (B220).
55 Zuckert, *Kant on Biology and Beauty*, 143.
56 Kant, *Critique of Judgment*, 211.
57 Ibid., 212.
58 This point is often obscured in Kant's argument.
59 Ibid., 212.
60 Zuckert extrapolated, "The Third Analogy might be taken to establish, then, that A and B are in community if I can perceive A then B, and then/also perceive B then A." Zuckert, *Kant on Biology and Beauty*, 139.
61 Kant, *Critique of Judgment*, 213.
62 Ibid., 213.
63 Ibid., 216.
64 Ibid., 372.
65 Ibid., 351. Zammito, "Teleology Then and Now," 216.
66 Kant, *Critique of Judgment*, 215.
67 Ibid.
68 Edmund Husserl, *Erste Philosophie (1923/4), Erste Teil: Kritische Ideengeschichte. Husserliana VII*, ed. Rudolf Boehm (The Hague, Netherlands: Martinus Nijhoff, 1956), 386.
69 For analysis and commentary on some of the differences between Kantian and phenomenological approaches, see Julia Jansen "Transcendental Philosophy and the Problem of Necessity in a Contingent World," *Metodo: International Studies in Phenomenology and Philosophy* 1 (2015): 47–80; Thomas J. Nenon "Some differences between Kant's and Husserl's conceptions of transcendental philosophy," *Continental Philosophy Review* 41, (2008): 427–439; Roman Ingarden, "*A Priori* Knowledge in Kant vs. *A Priori* Knowledge in Husserl," *Dialectics and Humanism* 1, no. 1 (1973): 5–18.
70 Edmund Husserl, *Philosophie der Arithmetik. Mit ergänzenden Texten (1890-1901), Husserliana XII*, ed. Lothar Eley (The Hague, Netherlands: Martinus Nijhoff, 1970), 40.
71 Ibid., 41.
72 Ibid., 42.
73 Edmund Husserl, *Logische Untersuchungen. Zweiter Teil. Untersuchungen zur Phänomenologie und Theorie der Erkenntnis. In zwei Bänden. Husserliana XIX*, ed. Ursula Panzer (The Hague, Netherlands: Martinus Nijhoff, 1984), 393; *Ideen zur einer reinen Phänomenologie und phänomenologischen Philosophie. Drittes Buch:*

Die Phänomenologie und die Fundamente der Wissenschaften, Husserliana V, ed. Marly Biemel (The Hague, Netherlands: Martinus Nijhoff, 1971), §13.
74 Aristotle, *Physics*, II, 3; *Metaphysics*, V, 2.
75 Roger W. Sperry, *Science and Moral Priority: Merging Mind, Brain, and Human Values* (New York: Columbia University Press, 1983). See also Mark A. Bedau "Downward Causation and Autonomy in Weak Emergence," *Principia* 6 (2003): 5–50.
76 Ernst Mayr, "Cause and Effect in Biology: Kinds of Causes, Predictability, and Teleology are Viewed by a Practicing Biologist," *Science* 134, no. 3489 (Nov. 10 1961): 1502.
77 Robert H. MacArthur, "Population Ecology of Some Warblers of Northeastern Coniferous Forests," *Ecology* 39, no. 4 (Oct. 1958): 599–619.
78 Mayr, "Cause and Effect in Biology," 1503.
79 Ibid., 1502.
80 Mayr, "Cause and Effect in Biology," 1503.
81 See Ronald A. Fischer's formulation of the fundamental theorem of natural selection, "The rate of increase in fitness of any organism at any time is equal to its genetic variance in fitness at that time." *The Genetical Theory of Natural Selection: A Complete Variorum Edition*, ed. J. Henry Bennett (Oxford: Oxford University Press, 1999), 35.
82 Mayr, "Cause and Effect in Biology," 1503.

Bibliography

Aristotle. *The Complete Works of Aristotle: The Revised Oxford Translation*, edited by Jonathan Barnes. Princeton: Princeton University Press, 1984.
Bedau, Mark A. "Downward Causation and Autonomy in Weak Emergence." *Principia* 6 (2003): 5–50.
Breitenbach, Angela. "Two Views on Nature: A Solution to Kant's Antinomy of Mechanism and Teleology." *British Journal for the History of Philosophy* 16 (2008): 351–369.
Cohen, Alix A. "Kant's Antinomy of Reflective Judgment: A Re-Evaluation." *Teorema: Revista Internacional de Filosofia* 23, no. 1 (2004): 183–197.
Fischer, Ronald A. *The Genetical Theory of Natural Selection: A Complete Variorum Edition*, edited by J. Henry Bennett. Oxford: Oxford University Press, 1999.
Friedman, Michael. *Kant and the Exact Sciences*. Cambridge: Harvard University Press, 1998.
Ginsborg, Hannah. "Kant on Understanding Organisms as Natural Purposes." In *Kant and the Sciences*, edited by Eric Watkins, 231–258. Oxford: Oxford University Press, 2001.
Ginsborg, Hannah. *The Normativity of Nature: Essays on Kant's Critique of Judgement*. Oxford: Oxford University Press, 2015.
Husserl, Edmund. *Erste Philosophie (1923/4), Erste Teil: Kritische Ideengeschichte. Husserliana VII*, edited by Rudolf Boehm. The Hague, Netherlands: Martinus Nijhoff, 1956.
Husserl, Edmund. *Philosophie der Arithmetik, Mit ergänzenden Texten (1890–1901). Husserliana XII*, edited by Lothar Eley. The Hague, Netherlands: Martinus Nijhoff, 1970.
Husserl, Edmund. *Logische Untersuchungen, Zweiter Teil. Untersuchungen zur Phänomenologie und Theorie der Erkenntnis, In zwei Bänden. Husserliana XIX*, edited by Ursula Panzer. The Hague, Netherlands: Martinus Nijhoff, 1984.
Husserl, Edmund. *Ideen zur einer reinen Phänomenologie und phänomenologischen Philosophie. Drittes Buch: Die Phänomenologie und die Fundamente der Wissenschaften. Husserliana V*, edited by Marly Biemel. The Hague, Netherlands: Martinus Nijhoff, 1971.

Ingarden, Roman. "*A Priori* Knowledge in Kant vs. *A Priori* Knowledge in Husserl." *Dialectics and Humanism* 1, no. 1 (1973): 5–18.

Jansen, Julia. "Transcendental Philosophy and the Problem of Necessity in a Contingent World." *Metodo: International Studies in Phenomenology and Philosophy* 1 (2015): 47–80.

Kant, Immanuel. *Metaphysical Foundations of Natural Science*, translated by James Ellington. New York: Library of Liberal Arts, 1970.

Kant, Immanuel. *Critique of Judgment*, translated by Werner S. Pluhar. Indianapolis: Hackett, 1987.

Kant, Immanuel. *Kants gesammelte Schriften V, Akademie Ausgabe*. Berlin: de Gruyter, 1902.

Kant, Immanuel. "On the Use of Teleological Principles in Philosophy." In *Anthropology, History, and Education*, edited by Günter Zöller and Robert B. Louden, 192–218. Cambridge: Cambridge University Press, 2007.

MacArthur, Robert H. "Population Ecology of Some Warblers of Northeastern Coniferous Forests." *Ecology* 39, no. 4 (October 1958): 599–619.

Mayr, Ernst. "Cause and Effect in Biology: Kinds of Causes, Predictability, and Teleology are Viewed by a Practicing Biologist." *Science* 134, no. 3489 (November 10 1961): 1501–1506.

Mensch, Jennifer. *Kant's Organicism: Epigenesis and the Development of Critical Philosophy*. Chicago: University of Chicago Press, 2013.

Moreno, Alvaro and Mossio, Matteo. *Biological Autonomy: A Philosophical and Theoretical Inquiry*. Dordrecht: Springer, 2015.

Nenon, Thomas J. "Some Differences between Kant's and Husserl's Conceptions of Transcendental Philosophy." *Continental Philosophy Review* 41 (2008): 427–439.

Quarfood, Marcel. *Transcendental Idealism and the Organism: Essays on Kant*. Stockholm: Almqvist & Wiksell, 2004.

Sperry, Roger W. *Science and Moral Priority: Merging Mind, Brain, and Human Values*. New York: Columbia University Press, 1983.

van den Berg, Hein. *Kant on Proper Science: Biology in the Critical Philosophy and the Opus Postumum*. Dordrecht: Springer, 2014.

Walsh, Denis M. "Organisms as Natural Purposes: The Contemporary Evolutionary Perspective." *Studies in the History and Philosophy of Science* 37, no. 4 (2006): 771–791.

Zammito, John H. *The Genesis of Kant's Critique of Judgment*. Chicago: University of Chicago Press, 1992.

Zammito, John H. "Teleology Then and Now: The Question of Kant's Relevance for Contemporary Controversies Over Function in Biology." *Studies in the History and Philosophy of Science Part C* 37, no. 4 (2006): 748–770.

Zuckert, Rachel. *Kant on Beauty and Biology: An Interpretation of the Critique of Judgment*. Cambridge: Cambridge University Press, 2007.

4 Husserl's logic of fitness
Parts, wholes, and phenomenological necessity

4.1 Introduction

At the beginning of *Formal and Transcendental Logic*, Husserl characterizes the discipline of formal logic according to a threefold signification of the Greek term *logos*. In the first instance, logos signifies the "laying together" or "setting forth" in language and this sense of *logos* corresponds to the discipline that Husserl calls "pure logical grammar" or "pure apophantics" that comprise the set of *a priori* and formal laws that regulate the combination of meanings into new unified meanings. This sense of *logos* as the lawful regularity of linguistic conjunctive, disjunctive, and hypothetical relationships has been predominant in Aristotelian logic and developed by the medieval logicians. The second signification of *logos* concerns the mental content involved in linguistic expressions. The signification of *logos* as the "laying together" of mind and its ideal objects is studied in the logic of consistency or non-contradiction. This level of logic concerns the inference rules that govern the relations among propositions, e.g., the rules of inference that determine whether a conclusion follows from a set of premises. Husserl locates the study of consistency with mathematical logic and distinguishes it from the Aristotelian apophantic tradition according to different kinds of formalization. Whereas the formalization in apophantic logic remains tied to both notions of class and existence, the formalization of the algebra made possible in Franciscus Vieta's mathematical logic frees the notion of form to apply to anything whatsoever with an empty universality that leaves material determination arbitrary. Form in this sense of the new mathematics is abstract, but is nevertheless lawful according to the norms of consistency or non-contradiction. The third signification of *logos* concerns – in Husserl's words – "affair-complexes referred to in speaking" or "what is thought."[1] Whereas the second sense of *logos*-as-mind finds its ultimate expression in the consistent validity studied in mathematical logic, this third sense of *logos* is the concern of the logic of truth – a level of analysis that is concerned with the truth of conclusions, not merely their validity. In other words, Husserl's distinction between the logic of consistency and the logic of truth parallels the standard distinction between validity and soundness. The truth of conclusions (soundness) is concerned with the evidence in which premises are given. In Husserl's philosophical project generally speaking, this evidence is found in the fulfilling intention of a truthful premise.

It is in this third sense of *logos* and its corresponding level of logic that Husserl secured the unity of formal logic and formal ontology. Formal ontology is the systematic analysis of the formal structures and relations involved in the pre-reflective experience of objects, including the operations involving these forms and relations. Formal logic is the systematic analysis of these same forms, relations, and combinations as meaning-categories and argument forms. Husserl's logic of truth unifies formal ontology and formal logic in an identity that obtains between the ontological object forms and logical meaning forms. This congruence occurs through meaning forms that are teleologically ordered toward fulfillment in our recognition of object forms. When a fulfilling intention confirms a supposition, this justification results from an identity between the objects and meaning forms. It is in this way, then, that the mature Husserl understood the relationship between phenomenology and logic – phenomenological analysis of the intentional structures involved in the formation of the judgments and categories proper to the correlative notions of formal ontology and formal logic. In other words, Husserl's mature development of transcendental logic – a phenomenological science concerned with the universal and necessary features of intentionality – that the correlative notions of formal ontology and formal logic are realized and obtain a common grounding with formal apophantics.[2]

This formal configuration of Husserl's mature philosophy of science has an original germination in Husserl's notion of pure logic in the *Logical Investigations*.[3] While the relations between logic, phenomenology, and ontology remain inchoate in these investigations, Husserl's analysis of parts and wholes provide an essential starting point for his mature approach. Through the clarification of parts and wholes in the Third Investigation, Husserl identifies methodological insights and principles that allow him to fix 1) his notion of phenomenological description of essential relations, and 2) his understanding of *a priori* necessity. Husserl's part-whole logic not only applies to the study of objects in formal ontology, but provides fundamental analytic and methodological principles that apply to his philosophical project in general. I submit that these principles can contribute to the development of a phenomenological account of biological form that is characterized by a logic of habitat fitness.

This investigation reconstructs Husserl's accounts of unified definite manifolds and part-whole logic and applies them to a phenomenological logic of habitat fitness. Husserl's logic of fitness opens a critical space in which multi-level habitat organizations can be identified and clarified in a way that makes ecologically emergent properties intelligible. The relationships among these complex systems are ultimately rooted in ontological assumptions concerning the relations between small-scale organization (parts) and large-scale organization (wholes) that are uncritically presupposed by computational models that define co-variance in merely informational terms. Robert H. MacArthur stated the problem of habitat fitness through relations among ecological parts and wholes,

> A great deal has been written about 'the whole being greater than the sum of its parts.' The trouble with this is that 'greater' and 'sum' are not appropriate words for analyzing wholes and parts. Most scientists believe that the properties of the whole are a consequence of the behavior and interactions

of the components. This is not to say that the way to understand the whole is always to begin with the parts. We may reveal patterns in the whole that are not evident at all in its separate parts.[4]

The phenomenological neutralization of the metaphysical assumptions concerning parts and wholes that follows in this investigation addresses the problem of habitat fitness through an account of changes that are described in terms of local co-variance of the population, community, and landscape levels. Multi-level explanations find their intelligibility, I argue, through the co-variance between levels that operates according to the lawful regularity of the necessary supplementation involved in alteration.

The problem of necessity in a phenomenological logic of habitat fitness concerns the identification and clarification of the kinds of invariant unities that can be known amidst the variant manifolds of a contingent world. This problem can be illustrated through a general reference to the Copernican revolution and the historical shift from a geocentric understanding of solar system variation to a heliocentric view. From a geocentric approach, the apparent motion of heavenly bodies, e.g., sun, moon, and stars, are accounted for in terms of the actual motion of heavenly bodies. By contrast, a heliocentric approach accounts for the apparent motion of heavenly bodies in terms of the motion of the observer. The observer undergoes variations that condition the knowledge of the variations in the world. In this historical sense, the problem of necessity is the problem of how to account for invariance amidst complex variation. The difficulty lies in the authentic clarification of how the variation in a contingent manifold is given as a unified determinate object with a necessary invariance-amidst-variability. Julia Jansen has recently formulated the problem of necessity in terms of the way in which a contingent sensible manifold is unified; "While the world is given as a sensible manifold, and, insofar as it is given, it is always contingent, its unity may carry and be *known* to carry necessity, in which case the unified manifold attains objectivity."[5] The necessary unity proper to sensible manifolds illustrates an important difference in the way in which Kant and Husserl accounted for the kinds of necessity in which sensible manifolds are unified as objects. While Kant's account maintains that a unified manifold receives the necessity of its unity in an asymmetrical relation to the transcendental unity of apperception, Husserl's symmetrical account recognizes that objects are passively synthesized with a necessary unity proper to themselves.

4.2 The problem of necessity and pure logic

Husserl's treatment of the problem of necessity in the *Logical Investigation* arises in the context of his discussion of propositional truths of scientific knowledge. Scientific knowledge is grounded in explanatory laws that provide justification for knowledge,

> To know the ground of anything means to see the necessity of its being so and so. Necessity as an objective predicate of truth (which is then called a

necessary truth) is tantamount to the law-governed validity of the state of affairs in question.[6]

Husserl thus identifies the necessity of a truth with justified propositional knowledge and makes the distinction between the necessity involved in individual and general truths. Individual truths contain propositions that indicate the actual existence of factual singulars and, as such, are contingent. By contrast, general truths involve assertions that infer the possible existence of individual facts from general laws. General truths yield justified propositions through deduction and have a necessity that results from the interconnectedness of basic laws that comprise a systematic theory.

The necessity involved in general truths is not vague and indeterminate, but obtains its determination in correlation with a given field of knowledge that comprises a determinate manifold, "The objective correlate of the concept of a possible theory…is known in mathematical circles as a manifold."[7] A manifold is a multiplicity in the formal and mathematical sense and Husserl's conception of the objective correlate of theoretical investigation finds its analytic expression as a theory of manifolds. It was Gottfried Wilhelm Leibniz who first envisioned the ideal of logic as "a discipline of mathematical form and strictness" and the breadth of logic to include "a mathematical theory of probabilities."[8] Leibniz's notion of *mathesis universalis* was the first systematic attempt to unify the formal apophantics of Aristotle with the mathematical logic that historically developed after Vieta. As Husserl states, "Our relation to him [Leibniz] is relatively of the closest" regarding the understanding of the relationship between logic and mathematics, a relation that finds its mature expression in *Formal and Transcendental Logic.*[9] In the *Logical Investigations*, Husserl identifies Leibniz's notion of a pure theory of manifolds with pure logic – a discipline that investigates the necessary unities of synthesized manifolds.

Pure logic is a discipline that underlies the methodological and normative features of logic in general and can serve as a formal theory of scientific reasoning. The discipline of pure logic identifies and clarifies meaning-categories (*Bedeutungskategorien*) and object-categories (*Gegenstandskategorien*) and the lawful combination of the two. For example, pure logic identifies and clarifies the laws governing combinations of meanings into propositions, arguments, and theories. The investigation into the meaning-categories yields formal concepts that correspond to the categories of formal ontology, e.g., concepts such as one, object, quality, relation, number, plurality, whole, and part. These formal categories are organized around the empty notion of "something" or "object as such" and are distinguished from the material concepts of object-categories. While the formal concepts apply to any object whatsoever, material concepts such as color, brightness, tone, intensity, plant, animal, and so on are organized around the highest regions among various given objects. The investigation into these object-categories yields a study of unified definite manifolds. To use the language from the middle period of Husserl's philosophy of science, these investigations yield "material" or "regional" ontologies.

4.3 Parts, wholes, and necessary fitness

Husserl's Third Investigation is a study of the lawful relations among two formal concepts of meaning-categories – the essential necessity among parts and wholes. He begins this investigation with the general distinction between simple and complex objects. Simple objects have no parts, while complex objects do. Complex objects, in other words, are wholes with parts. This distinction between wholes and parts that arises in complex objects is primitive, which is to say, no other terminology can provide a concept that is more basic nor can the relation between wholes and part be clarified by a more primitive terminology as long as parts and wholes are taken in general indeterminateness of formal concepts of meaning-categories.[10] Parts and wholes are fundamentally correlative, which is to say, they are defined in a mutual relation. Parts part wholes and wholes whole parts, that is, each is unintelligible without relation to the other.

The first distinction Husserl then makes concerns two types of parts – independent (*Selbstständigkeit*) parts (pieces) and non-independent parts (moments). Moments are parts that are non-independent from their wholes. Consider, for example, a material object taken as a whole – it cannot be presented without reference to its extension, surface, color, and brightness. Consider, in particular, a leaf from a fallen tree. These constituent parts of the leaf are organized through relations of dependence. The brightness of the leaf cannot be presented without presupposing its color, the color of the leaf cannot be presented without its surface, and its surface cannot be presented without its extension. These presentational moments of the leaf supplement each other and their whole necessarily, which is to say, the necessary supplementation involved in complex wholes defines Husserl's notion of moment (non-independent part).[11]

Pieces, by contrast, are independent parts that can by their nature be presented apart from the other parts forming a complex whole. Consider, for example, a tree as a whole comprised of branches, trunk, bark, roots, and so on. These parts can be presented separately – aspects of the tree branch can be presented in separation from the other independent parts of the tree, e.g., extension, weight, and shape. Not only can these parts be presented independently from the whole of the tree, e.g., the tree weight and branch weight have separable sums, but the parts can be independently presented from each other, e.g., the shape of the branches and trunk are not dependent on each other. Husserl recognized that certain aspects of a part are altered when removed from its whole while other aspects retain their identity. The fallen branch is separated from its functional unity with the whole tree and its corresponding functional sense is altered – the branch no longer functions in the nutritive dynamic of its whole – and therefore the functional sense of the branch has a non-independent relation (moment) to its whole. The fallen branch is a branch of the tree in name only (homonymously), which is to say, abstractly. However, Husserl's point is that the phenomenally presentational aspects of the branch – e.g., extension, weight, and shape – have senses that are independent in that they are not altered by their separation and are therefore not in need of supplementation. The independent branch can be presented apart from its functional incorporation into the whole tree and can be presented in its own unified right – as wood. Husserl thus defines a "piece" as any part that is independent relative to the whole of which it is a part.[12]

Husserl's mereology develops out of this basic distinction between independent and non-independent parts, a distinction that relies on a notion of necessity at work in the supplementation involved in alteration. As we will see later in this section, this notion of necessity among part-part relations and part-whole relations is basic to his presentational account of dependence and has implications for the way Husserl's conception of necessity relates to Kant's. The important point here is that the principle of relative dependency not only goes all the way down in Husserl's conception of part-part relations, but even extends to the basic part-whole relation itself. In other words, parts exist only in a relative dependency to wholes and while some parts can become presented as objects in their own right as wholes and thereby become a self-founding *concretum*, this nevertheless presupposes a more original complex whole.

Husserl's mereology proceeds from the distinction between (non)independent parts (pieces and moments) to founding-founded relations, relations that Husserl uses extensively not only in the Third Investigation, but in his later understanding of constitution.[13] Generally speaking, Husserl employs founding-founded relations extensively in his subsequent articulation of the network of definitions and laws governing the manifold of relationships that follow upon the piece-moment distinction. Husserl defines foundational moments as:

A) founded moment – a moment for which another moment provides a foundation in the formation of a whole; such that founded moment A supposes and forms a unity with moment B or whole W according to necessary association.

B) founding moment – provides a supposition and forms a unity with another moment and for the whole that it forms with its associated moments; such that moment B founds moment A such that B is the supposition of and unified with A or whole W according to necessary association.[14]

Consider a forest as an extended illustration of these founding-founded relations. The forest has a manifold of moments that can be considered in various founding-founded relationships. The populations of trees provide habitat for a manifold of insects, birds, and other animals. The set of interactions involved in this habitat relation could be considered a specific type of collective according to relations of foundation. If the habitat collective is merely a sum or aggregate with no founding-founded relations, then it is merely a whole in a rather wide sense, that is, determined merely in terms of abstracted unifying moment such as number or content. The habitat-as-whole would be merely a collective that lacks inherent organization. If the tree populations interact in a meaningful relation with the insects, birds, and other animals, this meaningful interaction can be expressed in various kinds of founding-founded relations. Consider, for example, interactions involving nutrition distribution. The tree provides nutrition for the insect, the insect is nutrient provision for the woodpecker, and so on. In this nutrient chain, one that has causal attributes as well, the nutrient provision of the woodpecker is founded on the insect, which in turn is founded on the beech tree. These founding-founded moments in the constitution of the nutrient chain could further be organized according to several additional distinctions that Husserl makes in his theory of foundation, e.g., immediate or mediate, remote or proximate, and so on. We could say, for example, that the beech tree is a mediate founding moment to the nutrient provision of the woodpecker, while the insect is the immediate founding moment. It is according to founding relations such as this that the nutrient fitness involved in a habitat is not merely a sum or aggregate, but an organized and organizing collective of meaningful relations that, as we have seen, have suppositions and forms of unified contents that are proper to the kind of object that it is.

While Husserl's understanding of wholes is not as developed as his mereology, his analysis of wholes is nevertheless instructive for drawing out the extent to which *a priori* necessity pervades his part-whole logic. Husserl defines a pregnant or proper whole as a set of contents (parts) united by a single, although possibly complex, foundation without the help of additional, non-essential contents (parts).[15] Notice that this definition of whole makes use of the non-independency of founding-founded relations. Every content (part) comprised in a proper whole is foundationally related with every other part or content comprised by that same whole. Husserl is clear that the unity of foundational relationships in proper wholes is not an additional moment over and above this interconnected unity of moments. Rather the "unifying moment" (*Einheitsmoment*) of the whole is immanent to the founding relations. This means that the unity of proper wholes arises from and finds its justification in the need for supplementation according to the lawful regularity of its non-independent parts. Even in the most minimal sense, the necessity of coexistence is sufficient to produce the unity of a whole. Indeed, the whole is nothing other than the interconnected unity of founding-founded moments and this unity is nothing other than the necessary lawful interconnections of moments.

This brief introduction to Husserl's account of necessity allows for a more general characterization of his position. Complex objects are presented as identities-in-a-manifold that are organized in the part-whole relations. The principle of lawful necessity not only pervades Husserl's conception of parts, but his conception of wholes. As we have seen, Husserl provides a symmetrical notion of presentational dependence that operates with a notion of necessity that can be defined as *necessary supplementation involved in alteration*. The symmetry between parts and wholes in complex objects is absolute according to *a priori* necessity. This has not only been seen in Husserl's first distinction between parts as independent or non-independent according to necessary supplementation, but in his characterization of the unity of wholes in terms of the necessity at work in founding-founded relations. To put it differently, Husserl provides a symmetrical model of presentational dependency that finds its evidential justification, as I explore in the next section, in his conception of noesis-noema correlation.

4.4 Multi-level generalizations

The problem of necessity is also tied, more broadly, to the process of generalization from the unified definite manifolds of perceptual sense to gradient kinds of generalizations involved in conceptualization. The unified definite manifolds involved in complex wholes can be distinguished according to the kinds of synthesis involved in perception and conceptualization. First, the unities involved in the perception of synthesized manifolds yield an identity. This identity-synthesis involved in perception implies that objectivity is not reducible to the presentational phases of the intended perceptual object, but rather has the unification of an identical individual. In short, perceptual synthesis in a manifold yields an identity – a unified and individual object.[16] Perceptual objects obtain their identity through the synthesis of appearances that manifest a unified individual.

Second, the unity involved in conceptualization is the result of a synthesis of a like with like (similarity). The synthesis involved in conceptualization is involved in the predications in which properties or attributes are related to objects, e.g., perceptual identities. For example, the judgment "The tree is tall" does not merely indicate that the tree has a certain vertical magnitude – that the tree is *this* tall. This would merely indicate that this tree is this tall and, as such, it is different than other tall trees. Instead, the judgment involves a predication of a property or attribute that involves a synthesis of like with like – that the tree is tall and as such is similar to other objects that have the attribute of being tall even amidst noticeably different heights. Such a predication involves the synthesis of similarity (like with like) at work in conceptualization.

The unities involved in the synthesis of like with like among collectives of objects can arise through a consideration of similarities or differences among the individuals of the collective. The collective is not merely an empty totality, but can be considered in terms of the similarity of objects with common attributes. The agreement of sense involved in this synthesis does not cancel the difference among spatially individuated objects. Rather the similarity of multiple objects is spread out in an array in which the differentiation of individuals persists. The unity here is the "unity of a plurality of kinship."[17] The persistence of the recognition of individuals in a collective distinguishes the identity of perceptual synthesis from the similarity of like with like involved in conceptualization.

The patterns of similarity involved in collectives can be analyzed at various levels of generality or universality. At one end of the spectrum is the lowest level of generalization that have only particular instances under them. Husserl calls the ideal objects of this level "eidetic singularities" that "manifest the lowest specific differences."[18] Second, the next level of generality focuses on the shared attributes of individuals of a collective and abstract a universal object-species. A species is a low-level generalization that abstracts the shared attributes of objects as identities and has a morphological essence that is determined in relation to eidetic singularities. At the third level of generalization are genera that abstract the common attributes among groups of species and individuals. We can grasp with lawful necessity the commonalities among species, e.g., beech, maple, and hemlock trees, and arrive at a genus – tree and plant in general. Finally, the highest level of generality involved in collectives of individual objects is a region. Husserl defines region as "the total highest generic unity" in which the genera and species of independently existing objects are organized, e.g., the regions of spatio-temporal materiality, animate organisms, and cultural achievements.

4.5 The distinction between analytic and synthetic necessity

The notion of necessity that is proper to the identity-in-a-manifold of object-categories and the synthesis of like with like involved in meaning-categories phenomenologically clarifies the authentic distinction between analytic and synthetic necessity. The degrees of increasing levels of generality progress according to an interest in the unity proper to the synthesis of "like with like" (similarity) and are determined according to the essential necessity of the eidetic sense of object-categories. The necessary supplementation involved in the alteration of presentational sense is also involved in both the determination of the various levels of generalization. We have seen that the lawful relations of formal concepts such as object, quality, relation, number, plurality, whole, and part have an essential difference from material concepts of object-categories, e.g., color, brightness, plant, tree and so on. Analytically necessary laws are operative in the formal concepts of meaning-categories that are independently founded on the indeterminate notion of "something" or "object as such." These analytically necessary laws are free from the determination of material concepts and the explicit or implicit assertion of individual existence. By contrast, synthetic necessity is determined by the material laws of object-categories and the specific nature of unified moments.

The essential distinction between formal categories and material spheres of essence provide the authentic basis for the distinction between analytic and synthetic propositions. An analytic necessity characterizes any proposition that is unconditionally universal in that it is free from the specifications of particular material content. The proposition "A whole has parts and parts have wholes," for example, has an analytic necessity that is merely determined by the formal concepts of whole and part. Analytic necessity can be distinguished from the specifications involved in material concepts when positing the relation among object-categories. Consider, again, the relationship between color and extension. Propositions regarding these material concepts do not involve an inherent relation to each other – color and extension have meanings that are independent from each other. Nevertheless, there is an essential necessity in the relation between color and extension. This necessity among material concepts of object-categories is synthetic in that the proposition "Color is not presented without extension" relates different object-categories that operate at different degrees of generalization. The relation between the sense of the concept of color and extension not only relate different propositional meanings but also relate varying levels of generalization – color is a lower-level generality than extension. Color does not analytically entail extension – analytic necessity is free of material determination of object-categories at whatever level of generalization. Synthetic necessity is defined, by contrast, as any law that identifies a founding relationship of material concepts through the clarification of the specific sense of unified moments. The synthetic or material *a priori* is the necessary lawfulness of the determinate sense in which various objects are disclosed.

4.6 Correlational *a priori* and intentionality

Husserl solidifies the conception of necessity that is operative in his notions of the *a priori* lawfulness through the logic of fitness in his theory of parts and wholes. We have seen that the necessary supplementation involved in alteration defines, in particular, Husserl's conception of a material *a priori* law as the essential relations of presentational dependence that organize unified definite manifolds. While Husserl's solution to the problem of necessity gets worked out in an eidetic analysis of the formal concepts of meaning-categories, he applies this solution extensively in his broader philosophical project, e.g., in the investigation into the logic of grammar and later theory of constitution. Husserl's mature theory of intentionality provides the ultimate justification for this conception of necessity. All consciousness is consciousness *of* such and such – or more formally – the experience of an object and the object of experience are internally related moments, not externally related pieces. All objects are complex objects insofar as they are given, minimally, to a subject.

Husserl characterized this internal unity between subject and object as mutually or reciprocally dependent moments of experience. This involves a departure from the distinction between "being for us" and "being in itself" in that an "object that is, but is not and in principle could not be an object of consciousness, is pure non-sense."[19] Husserl thus rejects Kant's distinction between the world as it appears and the world as it is in itself and shifts explanatory emphasis to the correlational variation involved in the determinate sense through which objects disclose themselves.

It is thus possible to speak of a "correlational *a priori*" proper to the noesis-noema correlate that characterizes the structures of intentionality.[20] For example, the variations involved in perception provide correlative unified definite manifolds – a manifold of appearances proper to the perceptual object and a manifold of perspectival orientation proper to an embodied noesis. The perceptual object, e.g., a tree, obtains its identity in a synthesis of a visual manifold that is generated by bodily variations, e.g., eye movements, neck movements, walking around the tree, and so on. The identity of the tree is manifested in and through the visual manifold. This is a significant point that can highlight another important difference between Husserl and Kant. For Husserl, the unity that is achieved in the synthesis identity of perception is inherent in the determinate sense of the object itself and not reducible to the perceptual achievements of the cognizing and embodied subject. The unity of identity is present in and through the manifold of contents such that the tree is *one* individual that is persistent throughout the manifold – the perceptual contents of the phases of the perceptual manifold necessarily supplement each other in an identity. The unity here is discovered (not achieved by the knower) and it is disclosed by and in the object in a "synthesis of coincidence" (*Deckungssynthese*).[21] The necessity here finds its justification in the lack of variation or alteration among the propositional attributes of the tree, e.g., extension, location, color, and so on. The object is a definite manifold that is unified as an identity – an individual whole – according to the essential necessity of its interrelated parts.

Kant, by contrast, maintained that the unity involved in objects of appearance is produced by the achievements of the cognizing knower. Kant's account of the unity of a sensible manifold reflects his separation between the faculties of understanding and sensibility. Unity presupposes synthesis that, in turn, is "an act of the [subject's] self-activity."[22] The unity that obtains through synthesis is an "act of the understanding" because unity, unlike form, demands the understanding. On the one hand, the unity of intuition is produced in the sensible synthesis and gives intuitions to objects. On the other hand, the unity of concepts, produced by intellectual synthesis, gives unified concepts to objects. In both cases, the unity is presupposed and conditioned by a "higher" unity, that is, the "original synthetic unity of apperception." The unity of apperception is necessarily valid and guarantees the possibility of self-consciousness, "The I think [cogito] must be capable of accompanying all my presentations."[23] This necessity is objective in that it unifies the intuitions of sensible manifolds. Indeed, Kant characterizes objectivity as "that in whose concept the manifold of a given intuition is united" and this unity is the achievement of the understanding, "that which itself is nothing more than the power to combine *a priori* and the bring the manifold of given intuitions under the unity of apperception."[24] This unity of apperception supervenes on objects in a sensible manifold as a one-sided relation between the subjective condition to the conditioned object.

The principle of necessity in Kant's conception of the *a priori* is asymmetrical (one-sided) from the point of view of Husserl's notion of the correlational *a priori* proper to intentionality. The correlational *a priori* operates with a symmetrical notion of necessity wherein the determinate senses of objects themselves obtain unity and fit together in states of affairs. As we have seen, this difference between Husserl and Kant's conception of *a priori* knowledge is particularly evident in their respective accounts of the synthesis of sensible manifolds. I submit that this difference has broad philosophical ramifications in general, and important implications for the philosophy of ecology in particular.

4.7 The *a priori* bound to the empirical and the problem of necessity

Husserl radicalizes and explodes Kant's notion of the *a priori* according to the various levels of generalization or universality. As we have seen, the formal objective *a priori* involves investigations into formal and mathematical logic and formal ontology. The material objective *a priori* is more complicated in that Husserl distinguishes between the *a priori* that is pure, exact, and "bound to the empirical."[25] Consider, in particular, the conception of necessity that is involved in the "*a priori* bound to the empirical." As we have seen, objects take shape as synthetic unities in the mode "they themselves" and not merely as appearances of objects. They have the determinate sense of their material content that is not merely unified according to the formal concepts of the meaning formations of objects in general. The material attributes of objects include, as we have seen, characteristics that operate at various degrees of generalization, e.g., species, genera, and region. In addition, these material attributes have particular determination as contingent matters of fact. It is in this sense that the material objective *a priori* is also a contingent *a priori*. The specific core of material content finds its determination in contrast to the specificity of different subsets of contents and is thus relationally limited in contingent matters of fact. Objects have relations with different objects and the extent of these differences limits the possible variations in which the object's specific attributes can be determined as an object of that type. The "*a priori* bound to the empirical" departs from the contents realized in empirical generalizations and intuits these contents as presumptively necessary for objects of a given type. It is a presumptive generalization with an empty necessity that is waiting to be filled out in that it has not tested the necessity of its relations and the universality of its generalizations. The "*a priori* bound to the empirical" can be distinguished from the "pure material *a priori*" that has achieved its fulfillment in the essential necessity manifest through a process of eidetic variation.

Even though the necessity-amidst-contingency involved in the "*a priori* bound to the empirical" has not yet been fully clarified in reflection, this does not imply that the necessity of unified manifolds is not proper to the objects themselves. Reflection on the way in which determinate objects in the world present themselves as sensible manifolds with necessary associations confirms that the world is pre-given with its own unity of coincidence (*Deckung*). As Husserl stated,

> there are breaks here and there, discordances; many a partial belief us crossed out and becomes a disbelief, many a doubt arises and remains unsolved for a time, and so forth. But ultimately, ... if the world gets an altered sense through many particular changes, there is a unity of synthesis in spite of such alterations running through the successive sequence of universal intending of the world – it is one in its particular details...it is in itself the same world. All of this seems very simple, and yet it is full of marvelous enigmas and gives rise to profound considerations.[26]

The necessary unities of sensible manifolds are not merely the result of subjective accomplishments, but have pre-given associations that are passively synthesized with a necessary unity proper to themselves.

Both Kant and Husserl share the conviction that transcendental philosophy attempts to identify and clarify the necessity of the lawful regularities in a contingent world through a reference to the necessary conditions of their knowability. They differ, however, in important ways with regard to their accounts of these necessary conditions. While Kant reasons from the necessary condition of the possibility of knowledge (unity of apperception) – the "I think" that accompanies all my representations – to that which is conditioned (unified manifolds), Husserl reasons from the conditioned (organized unity of definite manifolds) to the condition (the structures of intentionality). In other words, Kant accounts for the necessity proper to the unity and organization of manifolds in a one-sided relation to the subjective accomplishments of the knower. In contrast, Husserl accounts for the necessary unities of sense in terms of a two-sided relation of intentionality that is inclusive of lateral unities of coincidence.

4.8 Conclusion: the problem of ecological emergence

Husserl's part-whole logic, understood through his mature theory of intentionality, provides a useful point of entry into a variety of debates concerning the supervenience of so-called "emergent properties." As we have highlighted previously, Husserl provides a symmetrical model of ontological dependence, one that can be considered a viable contender in debates concerning the structure (morphological state) and functional dynamic of emergent properties. Indeed, Husserl's symmetrical approach cuts through the contemporary debates concerning downward causation by providing an alternative to the one-sided ontology of strong reductionism, on the one hand, and a dualistic ontology that maintains the inherent existence of wholes and parts respectively. From a Husserlian perspective, emergent properties are not merely illusory (as in the case of strong reductionism) nor do they have unities that are ontologically independent from their parts (as in dualist approaches). Phenomenology is neutral with regard to the metaphysical status of emergent properties and this neutrality is accomplished by a presentational model of ontological dependence that privileges the explanatory features involved in the covariance of supervenient relations. Phenomenological descriptions of essential necessities are also different from efficient causal explanations.[27] The remainder of this investigation provides a critique of some of the standard asymmetrical models and develops a logical space in which to address questions concerning principles of co-variance that are operative in multi-level explanations in ecology.

Questions concerning emergence revolve around the general notion that some large-scale systems acquire a degree of organizational complexity such that they exhibit novel properties that are autonomous from their constituent parts and cannot be predicted on the basis of the lawful organization of simpler systems.[28] This concept of emergence has enjoyed productive application in scientific analyses in fields ranging from cognitive science to molecular biology. Consider again, for example, the difference between an aggregate of trees and a forest. Both strong and weak approaches to emergence maintain that the forest is distinguishable from the mere aggregation of trees. To miss the forest for the trees is to miss the functional or meaningful patterns of organization proper to "forest" for an aggregate of individual trees merely organized according to collective identity relations.

One possible approach to the trees-forest relation is to start with the supposition that trees are ontologically basic in a way that parts are ontologically basic to wholes. The whole is nothing but the combination of the parts in that the whole is intelligible through the associations among parts and does not have intrinsic existence without them. On this strong reductionist view, the forest is a so-called "bottom-up whole" in the following respects:

1) The material of whole is nothing but the material of parts.
2) The state or property of the whole is nothing but the combination of parts and the parts themselves.
3) The dynamics (patterns) of states are nothing but the combination of parts.
4) The efficient causes of states and dynamics are nothing but the combinations of the parts.[29]

The predominant Nagelian framework that developed through the mid-20th century approached an understanding of question of emergence in bottom-up wholes according to the supposition that wholes are ontologically dependent on parts and have no causal autonomy of their own.[30] This strong reductionist approach gives explanatory privilege to the efficient causal relations among parts. The functional dynamics of the forest presuppose and form a causal unity with the causal features of the material properties of the trees. Efficient causality not only performs the exclusive explanatory work among part relations, but approaches the question of emergent attributes of wholes in terms of downward causation in multi-level organizations. The task in the Nagelian framework is to explain the complexity of large-scale organization in terms of the causal lawfulness of the small-scale or basal conditions.

The Nagelian model of reductionism, generally speaking, has faced two major challenges. The first challenge arose in the context of 20th century quantum mechanics and the explanatory difficulty accounting for the elementary particles of the quantum field in causal terms. In short, quantum particles do not appear to behave in predictable causal relations, that is, efficient causality is not enough of a fine-grained explanatory instrument to get at the patterns of alteration in elementary particles. This is especially acute when considering the lack of temporal duration among the cause-effect relations of elementary particles. The second challenge to the Nagelian model arose through a concern regarding whole-part causal relations. The Nagelian model is asymmetrical in that causal explanations that operate at the basal conditions are sufficient to explain the causal relations of large-scale organization. Complex causal patterns at the level of large-scale organization are, at most, epiphenomenal insofar as the complexity of these patterns has no unifying moment with other complex patterns of similar levels of organization. This causal asymmetry moreover had difficulty accounting for apparent downward causation from higher-level organizations to lower-level basal conditions.

These challenges have led to an increased interest in revising a strictly asymmetrical model of multi-level interaction to be able to entertain the possibility of a two-sided causal relation that operates among multi-level systems of complexity. A causal network operating in a large-scale organization, in this revision of the Nagelian model, can result in alterations of causal networks in the small-scale or basal conditions.[31] We find this kind of downward causation in a broad array of contexts. Consider, for example, the psychosomatic context of an episode of fear in which first personal presentational contents result in alterations in blood molecules. The first personal presentational content of perceiving a fearful dog, e.g., its large teeth and piercing bark, can be considered a complex whole. The first-person encounter with the fearful dog not only is resultant of causal networks of brain neuro-physiology, but can in turn result in changes in brain neuro-physiology and blood molecules. An episode of fear in an encounter with a fearful dog, in other words, can result in changes in lower levels – we have good experimental evidence indicating that the causal directions involved in the psychosomatics of fear are not asymmetrical, but that the coagulation of blood molecules causally results from the causal properties of the superveniencing encounter. Or consider again the example of the forest and the tree, this time through a revised part-whole logic that is willing to entertain the possibility of downward causality at the level of habitat. The habitat could be considered through the unity of causal networks of constituent parts, e.g., the trees, and the emergent causal networks proper to the forest. A non-reductive materialist approach that had logical space for emergent properties could examine the causal variance of the relative levels (basal and emergent). Consider again the causal networks involved in nutrient distribution among interspecific populations. The nutrition afforded to the woodpecker presupposes and forms a causal unity with insects, and the nutrition affordance to the insect, in turn, presupposes and forms a causal unity with the beech tree. The nutrient chain in woodpecker–insect–beech tree interactions can, at first glance, be explainable in terms of the strong reductionism of the classical Nagelian model – the nutrient distribution follows an asymmetrical causality from the lower to higher interspecific levels of interaction, e.g., beech tree, insect, woodpecker. But this asymmetrical approach becomes dissatisfying when presented with the theoretical demands for coherence. This demand arises, for example, from questions concerning the two-sided causal relations among interaction levels. We can ask with Roger Sperry how this approach accounts for the flight of bird.[32] When the bird takes flight, it not only brings its constituent parts with it, e.g., the organismic causal network of molecules involved in the woodpecker's own nutritional system, but extends the distribution interaction in an empty range of possibility in which the explanatory force of the causality operative in the nutrient chain become too fine-grained. Indeed, the woodpecker is a constituent part of this causal network with an organismic individuality of its own, one that introduces a self-reflexivity that secures it as a distinct explanatory level in the nutrient interaction. The large-scale organization operative at the level of the organism, e.g., the behavioral patterns involved in reproductive success or migration strategies, have a complexity that does not seem to be compressible to the lower levels. These incompressible causal networks, while causally dependent on the lower levels, introduce a level of complexity with an autonomy not reducible to the lower levels.

This is not mysterious. Jaegwon Kim calls this revision of the Nagelian framework – non-reductive materialism. Non-reductive materialism provides a logical space to consider emergent properties. Incompressible causal networks operative in large-scale organization have part-whole relations that – while they presuppose and form a causal unity with lower-level basal conditions – exhibit patterns of unification that are not necessarily entailed by lower-level causal explanations. There are three problems with non-reductive materialism from a symmetrical approach like Husserl's:

1) How is it that higher-level properties of large-scale organization, e.g., incompressible causal networks, alter the conditions from which they arise?
2) Does downward causation violate the causal closure of the microphysical or basal conditions? The issue of causal closure in multi-level interaction requires explanation.
3) What are the methodological principles that govern micro-level causal explanation with higher-level explanations? Micro-level causal explanations compete with higher-level causal explanations, rather than one-sided subordination.[33]

I take it that these questions – and the problems they indicate for standard models – are unresolved in a contemporary context. The dissatisfaction involved in these unresolved questions puts into relief the asymmetrical ontological assumptions of the standard model. When you start with a realist ontology of parts, the demands of theoretical coherency require you to give an explanation of how wholes emerge from parts. This theoretical requirement works the other way as well – if you start with a realist ontology of wholes, you eventually are required to give an explanation of how parts emerge from wholes. A symmetrical model like Husserl's, however, is metaphysically neutral when it comes to the real existence of parts or wholes but rather considers ontological dependency in presentational terms. As we have seen earlier in this section, this theoretical attitude shift is animated by an interest in the correlations among parts and wholes and, in particular, to the lawful regularity of necessary associations between, for example, micro-level parts and higher-level organizations. By taking questions of ontological dependency in terms of modal presentational contents, theoretical interest is freed from this metaphysical assumption (one-sided dualism) and secures the privilege of examining necessary relations as carrying the explanatory weight, the kind of necessary relations at work in the co-variance of multi-level explanations.

In conclusion, I would like to briefly sketch how Husserl's part-whole logic might help resolve one of the problems that face versions of non-reductive materialism, namely – the problem concerning the competing levels of causal explanation at the micro-level of basal conditions and large-scale organization. It seems as though there is a vicious circularity that does not meet the theoretical demands for coherence insofar as causal explanation of micro-level is inclusive of the downward causation of large-scale organizations. The justification for the causal network at the micro-level is supplemented by this two-sided causal interaction with the large-scale causal organization. The concern with downward causation in a multi-level framework assumes that the only relevant relations that obtain among levels is causal. Only brute efficient causality does the explanatory work.

But what if the relationship between large-scale organizations and micro-level basal conditions were understood in presentational terms oriented by an explanatory interest in broader ranges of necessity? This puts on the table not only the necessary relation between cause and effect, but the kinds of necessary relations involved in Husserl's model of founding-founded relations.[34] This model, as we have seen, construes ontological dependency in more fundamental terms of the lawful necessities of the supplementation involved in alteration. When a theoretical interest privileges this concern with the necessary alteration involved in supplementation in the examination of multi-level interaction, the issue becomes less about chasing one's causal tail, so to speak, but rather the identification of points of congruence among the alterations of the respective levels. This phenomenological move, in other words, is oriented by an interest in the points of co-variance among the multi-level interaction of complex systems. A variant property can be defined as a property that is altered through change. Co-variant properties, in turn, are identified in the alteration of two or more variables. The explanatory issue concerning multi-level organization, from this phenomenological perspective, concerns the co-variance between levels that occur according to the lawful regularity of the necessary supplementation involved in alteration.

Consider once more the example of the forest and the trees. Let's assume that the trees are merely an aggregate collection without moments of meaningful unity beyond their individual identity. The aggregate of trees is a collection of a manifold of individuals that share a common identity. But now let us assume that we can intelligibly talk about the forest as an organized interspecific collection with functional and emergent attributes such that these complex organizations are meaningful in a way that constitutes their own level of description, e.g., incompressible causal networks. What is the descriptive correlation between the concepts trees and forest? The tree-forest correlation is not a correlation of independent parts (pieces) comprising their separable and autonomous level of complexity (as in ontologically dualistic approaches). Rather, the tree and forest are dependent moments whose dependency lies not ontologically determined but determined according to the lawful regularity of necessary association. Do the trees come before the forest or does the forest come before the trees? Does upward causation fully explain incompressible causal networks, some of which have downward causal determinations at the micro-level? These are the circular questions that arise in asymmetrical substantialist approaches. By contrast, symmetrical approaches such as Husserl's privilege a presentational model of explanation that indexes the metaphysical prioritization of parts or wholes in favor of their relative dependency. Explanation of the tree-forest correlation tracks the necessary associative patterns that co-emerge in the correlation. The presentational sense of the trees as an aggregate collection is a founding moment in the presentational sense of the forest as an organized collection. The concept of trees does not logically exhaust the concept of forest, however. The forest has large-scale functional organization proper to a habitat – organized manifold of ecological fitness – that is not conceptually reducible to the trees. The forest-as-habitat has an incompressible pattern of necessary associations that, in principle, has its own determinate sense of meaningful contents. The process of explaining the relation between the aggregate of trees and the forest habitat occurs through the co-variance of the necessary associations that operate in each respective level. Explanatory inquiry follows the necessary structures of each level and plots inter-level points of convergence through moments of co-variance between the structures. The variance of the large-scale organization is coupled with the variance of the micro-level basal conditions to comprise correlative presentational structures with their own determinate structures. Multi-level explanation in ecology can occur, I suggest, through the co-variance between levels that operates according to the lawful regularity of the necessary supplementation involved in alteration.

Notes

1 Edmund Husserl, *Husserliana XVII Formale and transzendentale Logik. Versuch einer Kritik der logischen Vernunft*, ed. Paul Janssen (The Hague, Netherlands: Martinus Nijhoff, 1974), 18.

2 John J. Drummond, *Historical Dictionary of Husserl's Philosophy* (Lanham: The Scarecrow Press, 2008); Dieter Lohmar, *Edmund Husserls 'Formale und transzendentale Logik'* (Darmstadt: Wissenschaftliche Buchgesellschaft, 2000).

3 Edmund Husserl, *Husserliana XIX Logische Untersuchungen. Zweiter Teil. Untersuchungen zur Phänomenologie und Theorie der Erkenntnis. In zwei Bänden*, ed. Ursula Panzer. Halle: 1901; rev. ed. 1922 (The Hague, Netherlands: Martinus Nijhoff, 1984). Hereafter, Hua XIX.

4 Robert H. MacArthur, *Geographical Ecology: Patterns in the Distribution of Species* (Princeton: Princeton University Press, 1972), 154.

5 Julia Jansen, "Transcendental Philosophy and the Problem of Necessity in a Contingent World," *Metodo: International Studies in Phenomenology and Philosophy*, 1 (2015): 48.

6 Edmund Husserl. 1975. *Husserliana XVIII Logische Untersuchungen. Erster Teil. Prolegomena zur reinen Logik*, ed. Elmar Holenstein (The Hague: Martinus Nijhoff, 1975), §63. Hereafter, Hua. XVII.

7 Ibid., §36.

8 Hua. XVIII, §60.

9 Edmund Husserl, *Husserliana XVII, Formale and transzendentale Logik. Versuch einer Kritik der logischen Vernunft*, ed. Paul Janssen (The Hague: Martinus Nijhoff, 1974), §23. Husserl did not think that Leibniz's notion of *mathesis universalis* provided an adequate account of how the unity that brings together apophantic logic and mathematical logic in a single science is achieved.

10 In Peter Simons' pioneering attempt to formalize Husserl's mereology, he criticizes Husserl's seemingly arbitrary privilege of complex objects over simple objects. This criticism also arises in Simons' comment on Husserl's concept of a pregnant whole, "The pregnant whole for the foundation relation in question offers [Husserl] the promise of being neither too large nor too small. But this concept is itself defined in terms of the relation of individual foundation, as we shall see below, so it cannot be invoked without circularity. I do not believe that Husserl saw the threat of circularity here...." Peter Simons, *Philosophy and Logic in Central Europe from Bolzano to Tarski: Selected Essays* (Dordrecht: Kluwer Academic Publishers, 1992), 90. This criticism is correctly applied to the Third Investigation where Husserl does not provide this justification. However, it is ultimately Husserl's analysis of intentionality in the Sixth Investigation and eventual development of the phenomenological reduction that provide the justification for the individual foundation of complex objects and Husserl's concept of pregnant whole. Husserl methodologically establishes his theory

of intentionality between the first and second editions of the *Logical Investigations* through his non-psychologistic investigation of the relationship between consciousness and presentational contents (sense – *Sinn*). Husserl understands presentational contents to be necessarily intentional and intentionality to be necessarily content directed. See Edmund Husserl, 1984. *Husserliana XIX Logische Untersuchungen. Zweiter Teil. Untersuchungen zur Phänomenologie und Theorie der Erkenntnis*, ed. Ursula Panzer (The Hague: Martinus Nijhoff, 1984), Sixth Investigation, §48; *Husserliana III Ideen zu einer reinen Phänomenlogie und phänomenlogischen Philosophie. Erstes Buch: Allgemeine Einführung in die reine Phänomenologie*, ed. Walter Biemel (The Hague: Martinus Nijhoff, 1950), §2 and 88.

11 Hua. XIX, Third Investigation, §24. For a pioneering study of Husserl's part-whole logic, see Robert Sokolowski, "The Logic of Parts and Wholes in Husserl's Investigations," *Philosophy and Phenomenological Research* 28, no. 4 (1968): 537–553. For an approach to the relationship between Husserl's part-whole logic and broader methodology, see John Drummond, "Wholes, Parts, and Phenomenological Methodology," in *Edmund Husserl: Logische Untersuchungen*, ed. Verena Mayer (Berlin: Akademie Verlag, 2008), 105–122. For a critique of Sokolowski, see Jay Lampert, "Husserl's Theory of Parts and Wholes: The Dynamic of Individuating and Contextualizing Interpretation – *Übergehen, Abheben, Ergänzungsbedurftigkeit*," *Research in Phenomenology* 19, no. 1 (1989): 195–212. For an account that explores the relationship between Husserl and Heidegger in light of part-whole logic, see Einar Øverenget, "The Presence of Husserl's Theory of Wholes and Parts in Heidegger's Phenomenology," *Research in Phenomenology* 26, no. 1 (1996): 171–197. For a criticism of Husserl's part-whole logic through a notion of biological function, see Helena De Preester, "Part-Whole Metaphysics Underlying Issues of Internality/Externality," *Philosophica* 73 (2004): 27–50. For additional approaches, see Dieter Lohmar, *Edmund Husserls 'Formale und transzendentale Logik'* (Darmstadt: Wissenschaftliche Buchgesellschaft, 2000); Aron Gurwitsch, *Studies in Phenomenology and Psychology* (Evanston: Northwestern University Press, 1966); Aron Gurwitsch, *The Field of Consciousness* (Evanston: Northwestern University Press 1964); Thomas M. Seebohm, "Reflexion and Totality in the Philosophy of E. Husserl," *Journal of the British Society of Phenomenology* 4 (1973): 20–30; Kit Fine, "Part-whole," in *The Cambridge Companion to Husserl*, ed. Barry Smith and David Woodruff Smith (Cambridge: Cambridge University Press, 1995), 463–485; Dallas Willard, "The Theory of Wholes and Parts and Husserl's Explication of the Possibility of Knowledge in the *Logical Investigations*," in *Husserl's Logical Investigations Reconsidered*, ed. Denis Fisette (Dordrecht: Kluwer Academic Publishers, 2003), 163–182; John Drummond, *Historical Dictionary of Husserl's Philosophy* (Lanham: Scarecrow Press, 2008); and Dermot Moran and Joseph Cohen, *The Husserl Dictionary*, (New York: Bloomsbury, 2012).

12 Husserl, *Hua XIX*, Third Investigation, §17.

13 Edmund Husserl, *Experience and Judgment: Investigations in a Genealogy of Logic*, ed. Ludwig Landgrebe, trans. James S. Churchill and Karl Ameriks (Evanston: Northwestern University Press, 1973), §81. See also Robert Sokolowski, *The Formation of Husserl's Concept of Constitution* (Netherlands: Springer, 1970).

14 Husserl, *Hua XIX*, Third Investigation, §14; Drummond, *Historical Dictionary*, 82.

15 Husserl, *Hua XIX*, §21.

16 Husserl, *Experience and Judgment*, §81.

17 Ibid., 323.

18 Husserl, Hua III, §30. See also John Drummond, "Synthesis, Identity, and the *A Priori*," *Recherches husserliennes* 4 (1995): 27–51.

19 Edmund Husserl, *Husserliana II, Die Idee der Phänomenologie. Fünf Vorlesungen*, ed. Walter Biemel (The Hague: Martinus Nijhoff, 1973), 19ff.

20 Husserl, *Hua. XVII*, §72; *Hua. III*, §90; 1970. *The Crisis of European Sciences and Transcendental Phenomenology: An Introduction to Phenomenological Philosophy*, trans. David Carr, (Evanston: Northwestern University Press, 1970), §46.

21 Ibid., "The Origin of Geometry," 360.
22 Immanuel Kant, *Critique of Pure Reason*, trans. Norman Kemp Smith (New York: Macmillan and Company, 1965), B130.
23 Ibid., B131.
24 Ibid., B136.
25 Husserl, *Hua. II*, 374.
26 Edmund Husserl, *Husserliana XI Analysen zur passiven Synthesis. Aus Vorlesungs- und Forschungsmanuskripten, 1918–1926*, ed. Margot Fleischer (The Hague: Martinus Nijhoff, 1966), 101.
27 This phenomenological neutrality neither affirms nor denies the real existence of objects in the world but rather investigates how objects are presented – or better disclosed – with determinative sense. In this regard, it is preferable to speak of emergent attributes or relational patterns rather than "properties." This presentational or disclosure model is methodologically secured through Husserl's methodological principle of the phenomenological reduction.
28 Jaegwon Kim, "Making Sense of Emergence," *Philosophical Studies* 95 (1999): 3–36; *Supervenience and Mind: Selected Philosophical Essays* (Cambridge: Cambridge University Press, 1993); Mark A. Bedau and Paul Humphreys (eds.) *Emergence: Contemporary Readings in Philosophy and Science* (Cambridge: MIT Press, 2008).
29 See, for example, Aristotle, *Physics*, II, 3; *Metaphysics*, V, 2.
30 Ernst Nagel, *The Structure of Science: Problems in the Logic of Scientific Explanation* (New York: Harcourt, Brace, 1961).
31 Principle of downward causation – "To cause any property (except those at the very bottom level) to be instantiated, you must cause the basal conditions from which it arises (either as an emergent or as a resultant)." Kim, "Making Sense of Emergence," 24.
32 Roger W. Sperry, *Science and Moral Priority: Merging Mind, Brain, and Human Values* (New York: Columbia University Press, 1983). See also Mark A. Bedau, "Downward Causation and Autonomy in Weak Emergence," *Principia* 6 (2003): 5–50.
33 These are modifications of Kim's concern in "Making Sense of Emergence."
34 Edmund Husserl, *Husserliana XVI, Ding und Raum. Vorlesungen 1907*, ed. Ulrich Claesges (The Hague, Netherlands: Martinus Nijhoff, 1973), "On the Doctrine of the Levels of Givenness of Things," 341–345.

Bibliography

Bedau, Mark A. "Downward Causation and Autonomy in Weak Emergence." *Principia* 6 (2003): 5–50.
Bedau, Mark A. and Humphreys, Paul (eds.). *Emergence: Contemporary Readings in Philosophy and Science*. Cambridge: MIT Press, 2008.
Drummond, John. "Synthesis, Identity, and the *A Priori*." *Recherches husserliennes* 4 (1995): 27–51.
Drummond, John. *Historical Dictionary of Husserl's Philosophy*. Lanham: Scarecrow Press, 2008a.
Drummond, John. "Wholes, Parts, and Phenomenological Methodology." In *Edmund Husserl: Logische Untersuchungen*, edited by Verena Mayer, 105–122. Berlin: Akademie Verlag, 2008b.
Fine, Kit. "Part-Whole." In *The Cambridge Companion to Husserl*, edited by Barry Smith and David Woodruff Smith, 463–485. Cambridge: Cambridge University Press, 1995.
Gurwitsch, Aron. *The Field of Consciousness*. Evanston: Northwestern University Press, 1964.
Gurwitsch, Aron. *Studies in Phenomenology and Psychology*. Evanston: Northwestern University Press, 1966.

Husserl, Edmund. *Husserliana III Ideen zu einer reinen Phänomenlogie und phänomenlogischen Philosophie. Erstes Buch: Allgemeine Einführung in die reine Phänomenologie*, edited by Walter Biemel. The Hague: Martinus Nijhoff, 1950.

Husserl, Edmund. *Husserliana XI Analysen zur passiven Synthesis. Aus Vorlesungs- und Forschungsmanuskripten, 1918–1926*, edited by Margot Fleischer. The Hague: Martinus Nijhoff, 1966.

Husserl, Edmund. *The Crisis of European Sciences and Transcendental Phenomenology*, translated by David Carr. Evanston: Northwestern University Press, 1970.

Husserl, Edmund. *Husserliana XVII, Formale and transzendentale Logik. Versuch einer Kritik der logischen Vernunft*, edited by Paul Janssen. The Hague: Martinus Nijhoff, 1974.

Husserl, Edmund. *Husserliana XVIII Logische Untersuchungen. Erster Teil. Prolegomena zur reinen Logik*, edited by Elmar Holenstein. The Hague: Martinus Nijhoff, 1975.

Husserl, Edmund. *Husserliana XIX Logische Untersuchungen. Zweiter Teil. Untersuchungen zur Phänomenologie und Theorie der Erkenntnis*, edited by Ursula Panzer. The Hague: Martinus Nijhoff, 1984.

Jansen, Julia. "Transcendental Philosophy and the Problem of Necessity in a Contingent World." *Metodo: International Studies in Phenomenology and Philosophy* 1 (2015): 47–80.

Kim, Jaegwon. "Making Sense of Emergence." *Philosophical Studies* 95 (1999): 3–36.

Lampert, Jay. "Husserl's Theory of Parts and Wholes: The Dynamic of Individuating and Contextualizing Interpretation – *Übergehen, Abheben, Ergänzungsbedurftigkeit*." *Research in Phenomenology* 19, no. 1 (1989): 195–212.

Lohmar, Dieter. *Edmund Husserls 'Formale und transzendentale Logik'*. Darmstadt: Wissenschaftliche Buchgesellschaft, 2000.

Moran, Dermot and Cohen, Joseph. *The Husserl Dictionary*. New York: Bloomsbury, 2012.

Nagel, Ernest. *The Structure of Science: Problems in the Logic of Scientific Explanation*. New York: Harcourt, Brace and Company, 1961.

Øverenget, Einar. "The Presence of Husserl's Theory of Wholes and Parts in Heidegger's Phenomenology." *Research in Phenomenology* 26, no. 1 (1996): 171–197.

Preester, Helena De. "Part-Whole Metaphysics Underlying Issues of Internality/ Externality." *Philosophica* 73 (2004): 27–50.

Seebohm, Thomas M. "Reflexion and Totality in the Philosophy of E. Husserl." *Journal of the British Society of Phenomenology* 4 (1973): 20–30.

Simons, Peter. *Philosophy and Logic in Central Europe from Bolzano to Tarski: Selected Essays*. Dordrecht: Kluwer Academic Publishers, 1992.

Sokolowski, Robert. "The Logic of Parts and Wholes in Husserl's Investigations." *Philosophy and Phenomenological Research* 28, no. 4 (1968): 537–553.

Sokolowski, Robert. *The Formation of Husserl's Concept of Constitution*. The Hague: Springer, 1970.

Willard, Dallas. "The Theory of Wholes and Parts and Husserl's Explication of the Possibility of Knowledge in the *Logical Investigations*." In *Husserl's Logical Investigations Reconsidered*, edited by Denis Fisette, 163–182. Dordrecht: Kluwer Academic Publishers, 2003.

5 Environing places and geometric space

5.1 Introduction

This investigation explores several differences between phenomenological and Newtonian conceptions of space. There are two relevant theoretical motivations for this exploration. First, the differences between epistemological idealisms and epistemological realisms in the ecological sciences can be clarified through a contrast between the basic assumptions that underlie their respective accounts of spatial distribution. We have seen that the basic question of 19th century plant geography was a distributive question: how are plant associations geographically distributed? The answers to this question were reflective in the competing classifications of plant associations and conceptualization of plant formations, e.g., Alexander von Humboldt's "plant forms" and Eugen Warming's "growth forms." We have seen in a previous investigation that Humboldt's fundamental assumptions and explanatory principles were reflective in his compounded notion of plant form. The physiognomy of landscapes can be a Newtonian science in a variety of senses: 1) the earth's geography is spatially distributed in a homogeneous system of univocal locations, 2) latitude, longitude, elevation, and altitude can be measured with geometric exactitude, and 3) the lawful necessity of its explanations are obtained in a mathematical logic.[1] Humboldt's mathematization of geography illustrates a conception of spatial distribution that supposes geometric forms as the basic mark of necessity. I submit that this supposition is characteristic of epistemological idealisms in ecology more generally, e.g., the experimental design of Frederick Clements' quadrats or meter plots. In contrast, the conception of geographical distribution in physiographic classifications of plant associations operated with different fundamental suppositions and explanatory principles regarding spatial distribution. Plant associations are primarily distributed according to water variations in the soil which is primarily determined by variations in surface geology and topography. Plant associations are properly classified according to moisture tolerance of soil conditions that are primarily determined by geological and topographical conditions that are ideographic. The conception of spatial distribution in these physiographic accounts of plant associations, I submit, is a distribution of geographical places. These particular geographical places are ideographic – they have a historical and spatial singularity. The mark of necessity in physiographic explanations is not derived from a mathematical logic proper to geometric forms, but is derived from the identification of water variations as the basic variable proper to both plant physiological and topographical change. This logical necessity of physiographic explanations supposes a logic of habitats that is not mediated by the idealization proper to geometric form. This investigation is an attempt to identify and clarify the indirect idealization of geometric conceptions of space to the embodied habituation of particular places.

Second, this investigation of the fundamental assumptions and explanatory principles of the spatial distribution of habitats is motivated by a theoretical interest in the clarification of the boundaries or spatial limits of habitat associations. I submit that geometric conceptions of spatial distribution encounter difficulties in accounting for the boundaries of habitat associations due to an indirect idealization. That is, geometric conceptions of spatiality lose their logical authenticity through an uncritical abstraction from the embodied habituation involved in perception. Perceptual experience harbors unified definite manifolds that are spatially organized with necessary unities. The mark of necessity involved in this phenomenological approach to the spatiality of habitats is rooted in the perceptual sense of spatial associations. This is not a mathematical logic of spatiality, but a logic of the spatial sense of relative locations that does not uncritically abstract from the perceptual sense of embodied habituation. This more concrete conception of geographical distribution can more effectively account for the boundaries or spatial limits of habitats as environmental places.

What are some of the basic differences between the notions of environing places and geometric space? This investigation attempts to partially answer this question through the identification and clarification of the distinction of two specific conceptions of place and space. The conception of place that is operative in this investigation emphasizes the embodied aspects of being in a place, that is, those aspects of embodiment that contribute to the senses in which the first-person perspective is situated in a particular surrounding or environing world (*Umwelt*). I call this conception of place that emphasizes the bodily aspects of being in a place – embodied habituation. The conception of space, moreover, that is employed later in this section is also specific. There are many different conceptions of space and the one used in this investigation is a geometric space in the sense of a homogeneous system of empty, univocal locations that is organized according to Euclidean planes, e.g., Isaac Newton's conception of absolute space.

The differences between embodied habituation and geometric space can initially be indicated by calling attention to some of the peculiar ways in which a bodily first-person perspective is located and embedded in a concrete lived situation. Especially in the case of auditory and visual perception, objects and relations among objects are given in orientation to an embodied perceiver with a first personal "here." Moreover, the embodied aspects of a first-person perspective that include an awareness of this first personal location, e.g., proprioceptive and kinaesthetic sensations, make an essential contribution to the process of perception. This perceptual process also includes the manner in which objects are pre-given in relation to each other, e.g., auditory and visual objects take up positional associations among themselves and are not merely pre-given in orientation to the embodied perceiver, but to each other. These pre-given positional associations, in turn, become habituated and sedimented in the familiar contexts of everyday life, e.g., home, neighborhood, workplace, and so on, and inform the anticipations involved in perceptual processes.

The orientation awareness involved in environed embodiment can be distinguished from geometric space in three preliminary ways. First, while environing objects are given "in the flesh," geometric space abstracts from the material plenum involved in extension – geometric space is empty space. Second, geometric space has uniform units of position determined according to geometric planes, whereas the positional units involved in the orientation awareness of environed embodiment are determined by the internal unity (not external relation) of a perceptual "here" and environing world. Third, conceptions of geometric space arise from an abstraction proper to the idealization (not generalization or formalization) involved in geometric figure and algebraic symbolization, whereas the environing objects given in first-person perception are given in a "real actuality" that is not achieved as a result of idealization.

My basic argument is that embodied habituation has a one-sided founding relationship to geometrical space. More specifically, my claim is that the internal relation between pre-reflective bodily self-awareness and environing world has a founding relation with the concept of geometric space and that this relation is not reciprocal or mutual. A lived place is not a geometrical space in a variety of specific senses and the intelligibility of geometric space ultimately presupposes embodied habituation. My analysis begins with a contrast between Newtonian absolute space and the orientation awareness involved in kinaesthetic sensations and proceeds to identify and preliminarily clarify several continua (unified definite manifolds) involved in environed embodiment that contribute to awareness of bodily location and movement.

5.2 Newtonian absolute space and the critique of relative rotational movement

Classical mechanics is founded on a conception of space that is distinct from material plenum and a conception of time that passes uniformly and independently of changes in the world. In contrast to relative spaces and relative times, Newton distinguishes absolute space and absolute time in order to characterize the independence of spatial extension from material fullness, on the one hand, and sequential duration from material change, on the other.

Classical mechanics departed from Pre-Newtonian cosmologies that maintain the world is necessarily a material plenum and that the idea of empty space is ultimately incoherent. Space is not a real entity distinct from the material composition of things with an independent ontological status of its own, in this Pre-Newtonian view, but the result of an abstraction involved in collections of relational organization of individual bodies that comprise a material plenum. While Pre-Newtonian conceptions of motion did not reject the claim that extension is a necessary feature of material things, they approached the question of motion through an explanatory framework that was founded on the motions proper to relative locations.

Newton defined the true motion of a body to be its motion through absolute space and developed his definition of true motion in contrast to his relational account of motion.[2] He did not think that motion could be fully accounted for merely in terms of relation locations and this explanatory limitation comprises a central argument for Newton's identification and explanatory prioritization of absolute space. More specifically, at the beginning of his *Principia* in the "Scholium," Newton presents his conceptions of time, space, place, and motion through an initial distinction between ordinary common sense and mathematical cognition. In ordinary common life, these concepts are relative and apparent whereas they are conceived mathematically as absolute and true. Mathematical space is not only absolute and true, but is itself immovable and without relation to anything external.

One of the experiments that Newton cites as a demonstration of the argument that any adequate account of motion must include a notion of absolute space is the so-called "rotating bucket experiment." This experiment is designed by suspending a bucket through a long rope that is wound until the rope is tightly twisted. The bucket is then filled with water and when it is released, the rope unwinds and rotates the bucket and water. At first, the bucket rapidly accelerates while the water remains at rest relative to the experimenter. Newton takes these relative differences to indicate that the relative rotation of immediately adjacent bodies (bucket) need not necessarily produce centrifugal endeavor in the water to recede from the axis of relative rotation. As the experiment progresses and the bucket continues to rotate, the water gradually begins to rotate as well. As the water gradually accelerates its rotation, it climbs the sides of the bucket until the water rotation achieves a relative rest in relation to the rotating bucket. At this stage of the experiment, the water has achieved the maximum centrifugal endeavor to recede from the axis of rotation that is common to both the bucket and water. Newton concluded not only that centrifugal endeavor is not a necessary condition for the presence of relative rotation, but that it is not a sufficient condition for the presence of relative circular motion between the body (water) and its immediate surrounding (bucket). From this conclusion, Newton then drew the overall conclusion of his "argument from effects" – that true rotational movement cannot be defined merely in terms of the relative rotations of surrounding bodies. In order to account for relative rotational movements, it is necessary to appeal to an absolute space – a homogeneous system of univocal locations that remains immovable in relation to anything external and represents a necessary and sufficient condition for the explanation of relative motion.

The conception of absolute space in classical mechanics represents a decisive moment in the relationship between mathematics and physics after Galileo. This concept of absolute space as either a methodological form of measurement or an ontological supposition regarding real entities in the world is a continuation of the breakthrough of Galilean physics that has as its basic supposition that material bodies extended in space are fundamentally intelligible mathematically. The difficulty with this view involves what Husserl called the "indirect mathematization" and the "mathematical indexing" of sensible bodies.[3] Following the British empiricists, Husserl maintained that the primary attributes of material things are originally given in a full perceptual sense of material plena that is not directly presented through the idealization proper to geometric shape and algebraic symbolization. As Husserl stated, "The difficulty here lies in the fact that the material plena—the 'specific' sense-qualities—which concretely fill out the spatio-temporal shape-aspects of the world of bodies cannot, in their own gradations, be directly treated as are the shapes themselves."[4] Consider, for example, the tactile perception of a feather that includes gradients of givenness that are not fundamentally expressed through the idealization proper to geometric shape or algebraic symbolization. The tactile perception is a unified manifold of combinations of softness with various degrees of lightness which are, in turn, presentational contents that are not directly expressed mathematically. While the Galilean physics that finds its expression in Newtonian classical mechanics has its own regional validity and form of justification, this regional legitimacy does not extend to a pre-given life-world in which spatial extension is originally grasped together with material fullness. To substitute a mathematically indexed totality of spatio-temporal materiality for a pre-theoretical world pre-given in perception is to substitute a "garb of ideas" for the sensible bodies of the world given in the flesh. As Husserl stated,

> To be sure, everyday induction grew into induction according to scientific method, but that changes nothing of the essential meaning of the pregiven world as a horizon of all meaningful induction. It is this world that we find to be the world of all known and unknown realities. To it, the world of actually experiencing intuition, belongs the form of space-time together with all the bodily [*körperlich*] shapes incorporated in it; it is in this world that we ourselves live, in accord with our bodily [*leiblich*], personal way of being. But here we find nothing of geometrical idealities, no geometrical space or mathematical time with all their shapes. This is an important remark, even though it is so trivial.[5]

The lived bodily comportment in a pre-given environing world in which the real actuality of material things is originally given in perception is pre-geometrical in the sense that it is not presented in a mediation with the idealization proper to shape and figure. This perceptual presentational content – given in the flesh – remains the concealed or hidden presupposition of a mathematization of nature that is granted methodological or even metaphysical privilege. What remains unacknowledged in this presupposition concerns the first order idealization of shape and figure. This presupposition is hidden or unacknowledged as the mathematical tradition gets passed on from one generation to the next through a process of sedimented idealizations that are not authentically renewed again and again. Husserl's phenomenological re-discovery of the lived body occurred through an attempt to make authentic the sedimented idealizations involved in mathematical conceptions of space. The result of this authentication, moreover, is that an account of orientation awareness and motion that appeals to the relative positions of moving things, the relative location of the perceiver, and the overall environing situation is necessary and sufficient to account for motion without having to postulate the absolute space of classical mechanics.

5.3 Pre-reflective self-awareness of the lived body

The distinction between the lived body (*Leib*) – the body as embodied in a first-person perspective – and the objective, physical body (*Körper*) – the body as an object through a first- or second-person perspective – plays an important role in a phenomenological account of spatial awareness and motion. The first-person perspective is irreducibly located in the situation of the lived body – which generally is defined not merely in terms of perceptual location, but practical contexts and encounters with others. An embodied and embedded point of view is not a view from absolutely nowhere, but a view from an absolute here. The phenomenological claim is not merely that the lived body is objectively embodied and located "here" but that embodiment makes an essential contribution to the process of perceiving and enacting motion.[6]

Consider, for example, making a sandwich in a kitchen. The embodied and embedded features of this practical performance involve a series of interrelated performances, e.g., obtaining the ingredients from the refrigerator, slicing the bread, applying the ingredients, and so forth. In reaching for the ingredient, I am pre-reflectively aware of the proprioceptive, kinaesthetic, and tactile sensations of my lived embodiment. First, proprioception concerns the sense of the relative position of adjacent bodily components and the tension (strength of effort) involved in movement. As I grasp the ingredient from the shelf of the refrigerator, I am pre-reflectively aware of my overall posture and the orientation relations among my shoulder, elbow, and wrist. My awareness of bodily positional schema is physiologically provided by skeletal striated muscles, tendons, and joint capsules and is pre-reflectively given through various combinations of muscle and tendon extension, contraction, and release. The sensory givenness of bodily posture is a continuum (unified definite manifold) that is integrated with the vestibular system in the awareness of bodily posture in relation to the weight of gravitational pull. When I reach for the ingredient, I shift my weight to the tips of my toes and feel the stretch along the continuum of extended arm movement.

Second, pre-reflective bodily self-awareness is provided through kinaesthetic sensations involved in bodily movement. While the terminology is not always fixed in phenomenological discourse, I take kinaesthetic sensations to be real (*reell*) contents that present movement of the lived body and sense organs. For example, in walking over to the counter, I have an awareness of the motion of my body, e.g., eye, neck, and leg movements. Like proprioceptive sensations, kinaesthetic sensations are a continuum of presentational content of movement that is integrated with orientational awareness. Kinaesthetic sensations could be considered in a founded relation with proprioceptive and vestibular awareness, which is to say that without awareness of bodily posture and weight, coordinated bodily movement is difficult to imagine. Consider, for example, these founding relations involved in ambulatory movement. The kinaesthetic sensations involved in walking over to the counter and slicing bread are integrated (form a unity) with the proprioceptive givenness proper to posture and the vestibular givenness involved in the balance achieved through the shifting of weight. While founded on proprioceptive and vestibular awareness, kinaesthetic sensations are not reducible to them in that they are inclusive of bodily location in relation to situational objects, e.g., the neck, eye, and leg movements, that are involved in the perception of the bread knife on the counter.

Third, pre-reflective bodily self-awareness is also given through tactile sensations, e.g., in cutting the loaf of bread with a knife, I have tactile sensations of the smoothness of the knife handle and the coarse surface of the bread even though I do not thematically attend to them while making the sandwich. Moreover, I pre-reflectively feel the tug of my sweater that accompanies my performance of slicing the bread through tactile sensations associated with bodily self-awareness. In short, touch is a sensory system that functions in the perception of objects and pre-reflective self-awareness of the lived body.

Fourth, pre-reflective bodily self-awareness is given through feeling sensations involved in interoception such as hunger and thirst, sentient pain and pleasure, and the affective features involved in bodily motivation and fatigue. These feeling sensations arise from the lived body itself in an effort to achieve homeostasis in relation to an embodied situation, e.g., the stomach tension and increase of saliva involved in anticipation of eating the sandwich and thereby resolve the empty intentions involved in hunger.

Each of these features of bodily self-awareness more or less automatically contribute to normal embodied perception and action without conscious attention. They are pre-reflective in that they are not presentational in the same way as material object is presentational. They are given in non-objectifying intentions that contribute to the grasping of the sandwich as an object, that is, a material object with spatial extension. Pre-reflective bodily self-awareness is not a reflective form of self-consciousness that attentively inspects or reflectively introspects with regard to one's bodily state. Rather, it is a tacit or latent embodied awareness of myself in perception and action, one that contributes a field of activity and affectivity as possibilities for mobility and as a volitional "I can."

5.4 Kinaesthetic sensations and objectivity

Pre-reflective bodily self-awareness involved in kinaesthetic and tactile sensations plays an essential role in the awareness of relational location and movement through providing a continuum (unified definite manifold) of perceptual profiles in a series of appearances.[7] Husserl primarily refers to these sensations to address a specific question: what is it that allows us to combine several appearances to be that of one and the same object? What provides the capacity to perceive one and the same object in a succession of changing appearances? First, we could appeal to qualitative matching through the series of profiles, e.g., the color, shape, and size of the sandwich. While qualitative matching is a necessary condition for the synthesis involved in visual perception, it is not a sufficient one. Visual perception of the sandwich is accomplished through a synthesis of a manifold of profiles, a manifold that has a correlative relation to the synthesis of qualitative similarity. A second necessary condition for this synthesis is the manifold of bodily location in relation to, for example, the sandwich as I walk around the counter. Awareness of the perspectival absolute here of the series of profiles is given through kinaesthetic sensations – as I walk around the counter, I am pre-reflectively aware of my leg, neck, and eye movements. Kinaesthetic sensations thus generate a continuum – a unified definite manifold that is given in and through the variant profiles of the sandwich.

The kinaesthetic continuum that co-varies with qualitative matching is itself unified in a way that is not merely sensations of localized bodily movement, e.g., leg, neck, and eye, but an integrated horizon or field – what Husserl called the "kinaesthetic-sensual total system."[8] When Husserl metaphorically refers to the lived body as an "organ of perception" the implication was that pre-reflective sensory integration forms its own emergent continuum that is founded on the other sensory continua. While kinaesthetic sensations involved in pre-reflective bodily self-awareness are not given independently of the other sensory systems, they nevertheless distinctively co-function in the determination of an object through a manifold of appearances. It is *as if* the integration involved in kinaesthesis were another perceptual organ, which is to say, it is a functional and not an essential difference. As Husserl stated, "The lived body is constantly there functioning as an organ of perception; and here it is also, in itself, an entire system of compatibility and harmonizing organs of perception."[9] It is this continuum, given in pre-reflective bodily self-awareness, which provides an essential and evidential feature of visual and tactile profiles involved in the perception of a material object. A correlative sequence between 1) the kinaesthetic continuum and 2) the continuum of qualitative matching allows for the co-variance needed for the identity-in-a-manifold of the same object through a series of appearances. As we have seen, the kinaesthetic and tactile sensations generate a distinguishable continuum through which an identity is disclosed.

5.5 Kinaesthetic sensations and environmental situations

Kinaesthetic and tactile sensations also contribute to the awareness of relative location. The kinaesthetic continuum is not merely inclusive of bodily movements themselves, but bodily movements that are inclusive of bodily location given correlatively with a habituated situation. Co-functioning in kinaesthetic sensations are unthematic presentational contents regarding relative bodily location *vis-à-vis* the absolute here of the embodied perceiver. As I walk around the counter, I am kinaesthetically aware of my movement from $here_1$, $here_2$, and $here_n$. This manifold of past bodily locations is unified in combination with retentional phases and according to the location of the absolute here in primal impression. As Husserl stated, "Every kinaesthetic system has its null-position and its basic directions of modification starting from that position."[10] Accompanying the awareness of bodily movement, in other words, is an awareness of bodily location that is a component of the unthematic presentational content involved in kinaesthetic sensations. As we will see further later in this section, this continuum of unthematic presentational contents plays an essential role in a phenomenological account of spatial awareness and motion.

The sequence of bodily location given in and through the kinaesthetic sensations involved in walking is itself related to the variation involved in the overall environmental situation. By "environmental situation" I simply mean the overall totality of possible perception in a given bodily location. The environmental situation is inclusive, for example, of the inner and outer noematic horizons involved in perceptual presentational contents. In walking around the sandwich on the counter, I not only visually perceive its two-dimensional profile, but I apprehend the sandwich as having a backside and thereby the sandwich is disclosed with an inner horizon. Second, the sandwich is disclosed with a noematic outer horizon through which a relational orientation is presented in a manifold, e.g., the sandwich is on the counter, in front of the cabinet, beside the sink, and so forth. This noematic outer horizon is not merely a perceptual horizon that presents relational position, but a practical horizon through which concrete performances are achieved, e.g., the bread knife referentially implies the sliced bread which is, in turn, practically implied by the making and eating of the sandwich. The point is that environing object and states of affairs are co-given in horizons of meaningful associative syntheses with the noematic object and contribute to its disclosure.

While the overall environmental situation is inclusive of the patterns of association involved in inner and outer noematic horizons, it is not reducible to them. The overall habituated situation is not centered on the noematic object, but the embodied perceiver. The difference between noematic horizons and the overall environmental situation can initially be identified according to the difference of the continuum – while the noematic horizons indicate relational positions and practical associations in relation to the noematic object, the overall environing horizon indicates an indexical embodied location. Environing objects and states of affairs not only take up associative relational position among themselves, but these noematic horizons are likewise given in correlation with the embodied perceiver's absolute here and thereby attain a unified plurality of the similarity of orientation. The noematic object is thus constituted as an environing object insofar as it is disclosed with a continuum of similar orientation centered on the absolute here.

The difference between noematic and environing horizons can also be identified through the open and vague sense of totality of possible perception to be actualized at a given embodied moment. Consider an extended quotation from Husserl regarding this total possible perception, in which he states:

> that we continually have an experiencing consciousness in this life, but in connection to this in the widest parameters, an emptily presenting consciousness of an environing world—this is the accomplishment of unity out of a manifold, multifariously changing intentions, intuitive and non-intuitive intentions that are nonetheless concordant with one another: intentions that in their particularity coalesce to form concrete syntheses again and again. But these complex syntheses cannot remain isolated. All particular syntheses, through which things in perception, in memory, etc., are given, are surrounded by a general milieu of empty intentions being every newly awakened; and they do not float there in an isolated manner, but rather, are themselves synthetically intertwined with one another. For us the universal synthesis of harmonizing intentional syntheses corresponds to "the" world, and belonging to it is a universal belief-certainty.[11]

Environing objects and states of affairs are passively synthesized through horizons of referential implication that provide a unity of plurality involved in the similarity of presentational orientation. This environing harmony is a result of similarities, rather than identities, among environing objects and states of affairs and when an interest is awakened and drawn toward a prominent object, so too is a concrete similarity awakened and determined.[12] Husserl describes this harmonious agreement as a "resonance" which he defines as an "association according to similarity," which is to say, resonance is the associative similarity pertaining to the indicative relations among environing objects.

When one appearing object points to another, a basic degree of resonance adheres in the reference. This is not to say that all objects are necessarily similar to each other or that dissociation is not attributable to the relation among objects. To be sure, environing objects such as the sandwich, counter, and cabinet, for example, may not have similar shades of color and are not similar in that sense. Moreover, anticipations can be disappointed by the fulfillment of something otherwise than expected. Indeed, the referential implications within an environing situation can also be interpreted as a conflict of sense when the dissimilarity and diversity of objects are thematized. However, as Husserl pointed out, this awareness of conflict is founded on a more basic synthesis afforded by an environing world: "it is evident that conflict presupposes harmony, disappointment presupposes fulfillment....It pertains to the essence in general of being-otherwise, conflict, to presuppose a base of harmony."[13] The total perception of possible environing objects, in short, manifests a continuum of position that, while founded on the absolute here of the lived body, is not reducible to the respective orientations insofar as the similarity of orientation is inclusive of the noematic horizons of these objects. For example, the perception of the sandwich on the counter essentially includes the unthematic presentational contents that the sandwich on the counter is likewise in the kitchen of the house.

The overall unity proper to environing objects is given by the enworlding (*Verweltlichung*) function of an environing world.[14] The unthematic total perception of environing objects and states of affairs manifests itself as a unified totality of harmonious association that hangs together, so to speak, in a continuum that has a distinctive overall passive synthesis. While this function is not reducible to the embodied perceiver nor is it fully thematic in either pre-reflective or reflective experience, it nevertheless makes an essential contribution to the perception of environing objects and states of affairs as they come to prominence and affect us. The fulfilling intentions occur in a context of the indeterminate horizon that harbors latent associations even though they are no longer or not yet genuinely apprehended. They are given in and through a pre-given harmony of sense that is accomplished by the worldhood of an environing world, which is to say, the total harmony of sense is a unified definite manifold that co-functions in perception and action.

5.6 Motion and relational location

The internal unity between pre-reflective bodily self-awareness and an environing situation not only makes an essential contribution to the process of perception, but can also be employed in an account of motion that appeals to relational location (as opposed to geometric space). Not only do kinaesthetic and tactile sensations function in various continua involved in the perception of an identity through a manifold of appearances, but they also contribute to the apprehension of the kind of change involved in the locomotion of various objects in an environing situation. In the earlier example of making a sandwich in the kitchen, the sandwich is a stationary object that occupies a stationary position. As we have seen, the perceptual and practical features of the sandwich were disclosed in related continua that conditioned the appearance of the sandwich as one and the same stationary object. How are these related continua also relevant to the awareness of the motion of an object? Which continua proper to environed embodiment contribute to the apprehension of the locomotion of objects?

The first continuum that plays an essential role in the givenness of motion concerns the qualitative matching of the object. Consider, for example, a bird flying through a forest wherein the shape, size, and color of the bird is given in visual perception and the tone, pitch, and sequence of the bird's call is given in auditory perception. As the bird flies through the trees, the qualitative similarities across the progressive manifold of appearances generate a synthesis that necessarily contributes to the apprehension of the bird as one and the same object. As we have seen earlier, the passive synthesis associated with the qualitative matching according to similarity is a necessary but not sufficient condition for the givenness of the identity-in-a-manifold proper to the noematic object. Second, the manifold proper to qualitative matching is internally unified with the continuum of pre-reflective kinaesthetic sensations that generate the manifold of visual profiles of the object. The neck, eye, and torso movements involved in visually tracking the flying bird through the trees generate a manifold of perceptual profiles through which the synthesis of qualitative similarity manifests an identity proper to the noematic core – the bird as one and the same object.

Third, the continuum of kinaesthetic sensations also contributes an additional aspect of presentational content. Not only does this continuum generate a manifold of perceptual profiles proper to the apprehension of the noematic object, but it also contributes to the locational awareness of the embodied perceiver as "here." As highlighted earlier, this additional aspect of the continuum of kinaesthetic sensations manifest presentational content proper to the embodied perceiver's here and thereby environing objects obtain the unified plurality of the similarity of orientation. For example, as I track the flight of the bird, my pre-reflective awareness of oculomotor sensations contributes to the visual location of my eyes within a body schema that is "here." We have then, one continuum that functions in double aspects – A) the generation of the manifold of perceptual profiles of qualitative similarity, and B) a positional field that is indexed in orientation to the embodied perceiver's location.

The fourth continuum that is relevant to an account of motion is proper to the environmental situation. As we have seen, the overall environing horizon of total possible perception can be distinguished from noematic horizons and contributes an essential unity to the habituation involved in perceptual processes. As the bird flies through the forest, for example, its positional orientation takes up progressive positional association with the various trees – the tree is ahead, in front of, and behind the progressive trajectory of the bird's flight pattern. These positional orientations proper to the outer noematic horizon of the bird are a necessary but not sufficient condition to account for the unified perception of its motion. While the variation among perceived positional association is internally unified with the variation involved in the bodily movement given in kinaesthetic sensations, the unified location of the perceiver's "here" is correlated with the overall pre-given unity of the perceptual horizon in a way that contributes presentational content to an account of motion.

Does an environing world contribute essential presentational content to the perceptual awareness of locomotion? I suggest that it does. The preliminary justification for this suggestion concerns the internally unified relation between the embodied perceiver's "here" and the overall environmental situation. The absolute here of the embodied perceiver that indexes orientation awareness has as its correlation the overall environmental situation – a situation that not only contributes to the unified perceptual disclosure of objects as one and the same thing, but as a limit for relational positional association. The motion of the bird's flight does not merely occur in the foreground of the positional associations of the trees, but the trees that are given in an overall unified plurality involved in the similarity of orientation. It is this unified plurality – the forest as a complex whole, if you like – that correlatively indexes the manifold of perceptual profiles with its accompanying positional associations. The bird does not merely fly through a perspectival aggregation of trees, but the forest as a place – an overall horizon that is an indexical limit to positional association.

A natural objection arises at this point in the account of motion sketched earlier that raises several questions. Is it really necessary to take recourse to the fourth continuum involved in the environmental situation? Can motion be sufficiently accounted for only through recourse to the previous three continua? The explanatory principle involved in these questions is that, all things being equal, the simplest explanation is the best and that it is a virtue of any account to avoid unnecessary complications. The fourth continuum comprised of the passive synthesis involved in the overall environmental situation, according to this objection, is an example of such an unnecessary complication. In brief response, it is necessary to take recourse to the fourth continuum of presentational content because it provides an invariant reference for the system of positional association and thereby avoids a lurking infinite regress of relative positions. For example, an account of the motion of the bird's flight cannot fully be determined merely through recourse to the bird's change in relational position to the trees considered as an aggregate of positions. Rather, the bird flies through an organized and organizing field of positions – an environmental situation that provides an overall unity that is co-variant with the embodied location of the perceiver. In other words, the first and second continua (qualitative matching involved in a manifold of perceptual profiles) lack an invariant reference point proper to positional association to fully determine the moving object. This invariance – or better, co-variance with the perceiver's embodied location – is given through the unthematic totality of possible perception at any given moment and confers a definiteness to the manifold of positional association.

5.7 Idealization of objective space

A phenomenological account of motion need not appeal to the absolute space of classical mechanics. Newton's distinction, however, between the relative spaces of ordinary life and the idealized space of geometry finds its authenticity through an account of the one-sided founded relation of the concept of geometric space as a homogeneous system of univocal locations to environed embodiment. This means that the concept of geometric space presupposes and forms a unity with the complex intentional correlate between the lived body and its environing world, but this supposition and unity do not hold in a reciprocal founding relation. This section outlines the phenomenological idealization involved in the concept of geometric space through a contrast between a lived and objectively spatial body.

An environing world is centered on the lived body and environing objects and states of affairs are given in orientation to the absolute here. Other animate organisms take their own place within the common situation, the place that I live with them.[15] Their bodies are first perceived in their physical constitution and subsequently their movement indicates active spontaneity due to the absence of other movement that might result in a change in their position. I pair their movement with my own and appresent kinaesthetic sensations in them through empathy. I thus apprehend them minimally as animated organisms with pre-reflective lived bodies. In this apperception, I realize that their pre-reflective lived body also perceives from an "ownmost" located perspective, which is to say, they must also be oriented by their own irreducible absolute here. While retaining the absolute here functionally presented in my pre-reflective bodily self-awareness, I simultaneously appresent the absolute here of the other animate organism.[16]

The difference maintained between the absolute here of my pre-reflective bodily self-awareness and that of the other animate organism accounts for a special sense in which an environmental situation is shared. The two bodily locations are given differently in that the former is pre-reflective, and the latter is appresented in a higher order act. The situation is shared, not through the mere co-presence of our physical bodies, but through the tacit acknowledgment of the presence of the animated other's pre-reflective bodily self-awareness. For example, when a cat brushes against my leg and meows for attention, I am presented with an opportunity for kinaesthetic and tactile sensation. I can reach down and pet it, feeling the smoothness of its fur and so on. The cat purrs when I rub its neck and arches its back when I stroke it, thereby displaying signs of enjoyment in the reciprocal tactile sensations involved in the touch. I implicitly acknowledge that the cat has tactile sensations corresponding to its kinaesthetic sensations and appresent the cat's absolute here through empathy. I thus have a tacit recognition that my environmental situation is shared.

A shared environmental situation involves a modification of horizonal awareness. It is not only my lived body that is environed, but the same objects and states of affairs envelop the animate other's pre-reflective lived body. The here of the first-person perspective thus occurs in a manifold of bodily location. As Husserl states,

> With this is given simultaneously the fact that the environing world of the alien pure Ego is the same one as that of my pure Ego; and this means: the physical things of the environing world are unities of a higher level, constituted by way of interpretation (*Eindeutung*).[17]

This, then, is the next level of the formation of the spatial body, namely, when I reflectively abstract from my absolute here and not only consider the way my physical appearance is given to the animate other, but the positional sense in which my physical body is given to the animate other. "But from that here I can then consider even my own Body as a natural Object, i.e., from that 'here' my Body is 'there,' just as the other's Body is 'there' from my 'here,' is there at a point in Objective space."[18] The spatial body thus becomes constituted when the recognition of the absolute here of two or more lived bodies relativizes my absolute here. As Husserl states,

> It is as localized...that other subjects are there for us. To the compass of that which is appresented with the seen Body pertain also the systems of appearances in which an external world is given to these subjects...their place is given to us as a 'here,' in opposition to which everything else is 'there.' But with analogizing, which does not result in anything new over against the Ego, we have at the same time the other Body as 'there' and as identified with the Body-as-here phenomenon. I then have Objective movement in space...[19]

At this point two interrelated questions arise: does the constitution of the objective space necessarily involve an intersubjective relation to other lived bodies? Is it not possible to arrive at a conception of the spatial body on the basis of the continuum of retentional bodily locations? The constitution of objective space is necessarily mediated intersubjectively and it is not possible to constitute one's body in space on the basis of a solipsistic continuum of retentional bodily locations for this reason: the absolute here is relegated to primal impression. The "here" loses its absolute index once the transition to a retentional phase occurs and becomes "here" as retentional. In the temporal flow of the first-person perspective, the indexical character of the absolute here is confined to the instantaneous now of the moment. As Husserl stated, "Corresponding to the absolute Now as the zero-point of temporal orientation is the absolute Here as the zero-point of all spatial orientation."[20] While the elapsed absolute here (zero-point of all spatial orientation) along with its correlated environing objects comprise co-variant manifolds with an identity in difference, this identity is underdetermined by the indexical character of the absolute here. The continuum of retentional "heres" is a unified definite manifold, but its unity is absorbed by the indexical function of the absolute here of primal impression. As Husserl stated, "While the latter can in principle take on every and all orientation with the exception of a zero-position, one's own lived body, however, can only take on very limited variations in orientation, precisely because of its tie to the zero-point."[21] In short, the formation of the concept of objective space is necessarily intersubjective in that the radical momentary character of the absolute here subordinates the continuum of retentional phases of the absolute here of a solipsistic subject because it lacks the simultaneity afforded by an intersubjective relation.

In addition to this intersubjective mediation, the formation of the concept of geometric space involves a process of idealization proper to geometrical plane. Idealization is a kind of abstraction that yields abstract objects having a determinate material content. Unlike the abstraction involved in generalization, idealization involves similar objects arrayed in a progression toward a limit that is not itself realized in any individuals of the progression.[22] The limit does not adhere in progression, but has the difference of alterity – it exists on a different plane. The ideal object is constituted through a shift of attention to this ideal limit. Consider, for example, a Euclidean geometric figure such as a cube. The ideal limit "cube" is understood through a progression that extends beyond the manifold of actual box-like, three-dimensional volumes to the purely possible and thereby yields an *a priori* object – a cube – rather than an empirical generalization (the abstraction of an identity typifying the similarities of objects). The ideal object is constituted through when the universal is understood against the manifold or array and without an awareness of the arrayed manifold, there would be no genuine awareness of the ideal figure of a cube.

The process of idealization involved in the formation of objective space as a homogeneous system of univocal locations takes the basic directional orientations – up/down, side/side, forward/back – in progressions toward ideal limits proper to a geometric plane. These morphological directions are the arrayed manifolds from which three-dimensional geometrical planes are idealized. Consider, for example, the two dimensions involved in visual perception: height and length. I visually perceive the sandwich or cat in a two-dimensional field as an identity-in-a-manifold having height and depth in co-variance with ocularmotor and cephalomotor kinaesthetic sensations. The idealized horizontal axis (side/side) and the vertical axis (up/down) are constituted in relation to the zero-point of orientation located between the eyes. The zero-point of visual orientation has a correlated focal point that is the unthematic center of attention of accompanying visual horizons. This correlated zero-point and focal point are the basis for the differentiation within each axis and between axes and the limit case for the arrayed manifolds involved in height and length. The disclosure of depth is not as straightforward and involves the sensory integration with tactile and kinaesthetic sensations. Husserl made extensive use of the kinaesthetic sensations involved in walking in his account for this third dimension of geometrical figures and space.[23] Sufficient for the present purposes is the evidentially justified fact that depth can be accounted for through the combination tactile sensation and the variations of the size objects as they approach and recede.[24] These three orientations are the arrayed and intersecting manifolds that are idealized through the limit case proper to geometrical plane. The limit cases function as universals in contrast to the arrayed manifold of orientations and are together constituted as an ideal object of three-dimensional axes intersecting at right angles. The concept of geometric space is constituted along these lines.

5.8 Conclusion: on the limitations of objective space

Newton's distinction between the relative spaces of ordinary common life and mathematical space finds its phenomenological authenticity in the contrast between the lived body and spatial body (body as apprehended in a homogeneous system of univocal locations involved in a process of idealization). The distinction finds its legitimacy not in Newton's affirmation that empty space ontologically exists independently of material plenum and has methodological priority in the determination of motion. Rather, the distinction finds its authentic sense through the identification and clarification of the embodied habituation involved in pre-reflective bodily self-awareness and the formation of the concept of geometric space. As highlighted earlier, the distinction is comprised of a one-sided founding relation between environed bodily location and objective space – the former is the supposition and underlying unity that makes the latter possible. The relation is not a reciprocal dependency in that the spatial body can only be thematized in a reflective idealizing act – it cannot be lived pre-reflectively.

Jean-Paul Sartre and Maurice Merleau-Ponty corroborate and extend this point. For Sartre, space and the spatial body can only be constituted abstractly, they cannot be lived.[25] The body in itself (pre-reflective lived body) always surpasses itself by refusing an ipseity with its past positions. It always already occurs in a lived place. As Sartre states,

> The body is perpetually the surpassed. The body as a sensible center of reference is that beyond which I am in so far as I am immediately present to the

glass or to the table or to the distant tree which I perceive. Perception, in fact, can be accomplished only at the very place where the object is perceived and without distance.[26]

For Merleau-Ponty, as soon as I try to posit bodily space or intuit the spatial body, "I find nothing in it but intelligible space."[27] He continues with a rich description:

> And indeed its [the body's] spatiality is not, like that of external objects or like that of 'spatial sensations,' a spatiality of position, but a spatiality of situation...The word 'here' applied to my body does not refer to a determinate position in relation to other positions or to external co-ordinates, but the laying down of the first co-ordinates, the anchoring of the active body in an object, the situation of the body in the face of its tasks. Bodily space can be distinguished from external space and envelop its parts instead of spreading them out, because it is the darkness needed in the theatre to show up the performance, the background of somnolence or reserve of vague power against which the gesture and its aim stand out, the zone of not being in front of which precise beings, figures and points can come to light.[28]

Environed embodied habituation does not involve idealized coordinate dimensions and the proportion and congruence of magnitudes in univocal locations. It involves a process of manifestation that could perhaps succinctly and precisely be described as a *non-spatial spatializing*. The absolute here involved in pre-reflective bodily self-awareness resists and even refuses determination in mathematical space – it does not and indeed cannot be reflectively modified.

In conclusion, environed habituation has a one-sided founding relationship to conceptions of objective space in that the internal unity between pre-reflective bodily self-awareness and environing world is the supposition for and the underlying unity of geometric space. In turn, the process of idealization involved in geometric space necessarily does not reflectively modify bodily location but remains merely "intelligible space." This difference between pre-reflective bodily self-awareness and the idealization involved in geometric space, I suggest, is a basic difference between the notions of place and space.

Notes

1 Humboldt's elevation measurement of the Chimborazo summit (6,544 meters) deduced the base of the mountain at sea level according to Laplace's barometric formula. See Alexander von Humboldt and Aimé Bonpland, *Essay on the Geography of Plants*, ed. Stephen T. Jackson, trans. Sylvie Romanowski (Chicago: University of Chicago Press, 2009), 147.

2 Isaac Newton, *Philosophiae Naturalis Principia Mathematica*, 3rd ed., ed. Alexandre Koyré, Bernard Cohen, and Anne Witman (Cambridge, MA: Harvard University Press, 1972). See also Robert Rynasiewicz's "By Their Properties, Causes and Effects: Newton's Scholium on Time, Space, Place and Motion. Part I: The Text," *Studies in History and Philosophy of Science* 26 (1995): 133–153; "By Their Properties, Causes

and Effects: Newton's Scholium on Time, Space, Place and Motion. Part II: The Context," *Studies in History and Philosophy of Science* 26 (1995): 295–321.

3 Edmund Husserl, *The Crisis of European Sciences and Transcendental Phenomenology: An Introduction to Phenomenological Philosophy*, trans. David Carr (Evanston: Northwestern University Press, 1970), 37.

4 Husserl, *Crisis*, 33.

5 Husserl, *Crisis*, 50.

6 This claim is developed in Ulrich Claesges, *Edmund Husserls Theorie der Raumkonstitution* (The Hague: Martinus Nijhoff, 1964); John J. Drummond "On Seeing a Material Thing in Space. The Role of Kinaesthesis in Visual Perception," *Philosophy and Phenomenological Research* 40 (1979–1980): 19–32; Shaun Gallagher, "Body Schema and Intentionality," in *The Body and the Self*, ed. José Luis Bermúdez, Naomi Elian, Anthony Marcel (Cambridge: MIT Press, 1998), 225–244; Dan Zahavi "Husserl's Phenomenology of the Body," *Etudes phénoménologiques* 10 (1994): 63–84.

7 Edmund Husserl, *Husserliana XVI, Ding und Raum. Vorlesungen 1907*, ed. Ulrich Claesges (The Hague, Netherlands: Martinus Nijhoff, 1973), 44; English translation: *Thing and Space: Lectures of 1907*, trans. Richard Rojcewicz (Dordrecht: Springer, 1998), 38.

8 Husserl, *Crisis*, 106.

9 Edmund Husserl, *Husserliana XI Analysen zur passiven Synthesis. Aus Vorlesungs- und Forschungsmanuskripten, 1918-1926*, ed. Margot Fleischer (The Hague, Netherlands: Martinus Nijhoff, 1966), 13; English translation: *Analyses Concerning Passive and Active Synthesis: Lectures on Transcendental Logic*, trans. Anthony Steinbock (Dordrecht: Kluwer, 2001), 50; c.f. Husserl, *Husserliana XVI*, 171, 144.

10 Husserl, *Husserliana XVI*, 303, 261.

11 Husserl, *Husserliana XI*, 101, 145. See also Aron Gurwitsch, *The Field of Consciousness* (Pittsburgh: Duquensne University Press, 1964); *Marginal Consciousness*, ed. Lester Embree, (Athens: Ohio University Press, 1985); and *Phenomenology and the Theory of Science*, ed. Lester Embree (Evanston: Northwestern University Press, 1974).

12 Husserl, *Husserliana XI*, 407, 506.

13 Husserl, *Husserliana XVI*, 97, 82; c.f. *Husserliana XI*, 407, 507.

14 Husserl characterizes an environing world as "the bare mobile core of the world." *Husserliana XXXII, Natur und Geist: Vorlesungen Sommersemester 1927*, ed. Michael Weiler (Dordrecht: Kluwer, 2001), 200. He similarly states, "The world unnoticed in its being-in-itself, in accordance with the sense of this being-in-itself is thereby given as my environing world." *Husserliana Materialienband IV, Natur und Geist. Vorlesungen Sommersemester 1919*, ed. Michael Weiler (Dordrecht: Kluwer, 2002), 224. Finally, "That which has already resulted for us in this regard is the fundamental basis of the original source of the determination of the concept 'world,' namely, the necessary form of the correlation 'I and my environing world,' and the correlation 'I and my pre-theoretical environing world.'" *Husserliana Materialienband IV*, 228. This point is developed further in my "The Environed Body: The Lived Situation of Perceptual and Instinctual Embodiment," *Studia Phaenomenologica* 12 (2012): 289–308; and "Embodiment and *Umwelt*: A Phenomenological Approach," *Social Imaginaries* 1(2) (2015): 53–72.

15 Husserl, *Husserliana IX*, 231, 177.

16 Husserl, *Husserliana V*, 109, 94.

17 Husserl, *Husserliana V*, 109, 95.

18 Husserl, *Husserliana IV*, 169, 177.

19 Husserl, *Husserliana IV*, 168, 176.

20 Husserl, *Husserliana XI*, 297, 584.

21 Husserl, *Husserliana XI*, 298, 585.

22 John J. Drummond, *Historical Dictionary of Husserl's Philosophy* (Lanham: Scarecrow Press, 2008), 101.

23 Husserl, *Husserliana XVI*, 336, 288.
24 Husserl, *Husserliana XVI*, 267, 310.
25 Jean-Paul Sartre, *Being and Nothingness*, trans. Hazel E. Barnes (New York: Simon and Schuster, 1984), 405.
26 Sartre, *Being and Nothingness*, 429.
27 Maurice Merleau-Ponty, *Phenomenology of Perception*, trans. Colin Smith, 2nd ed. (New York: Routledge, 1958), 117.
28 Merleau-Ponty, *Phenomenology of Perception*, 115.

Bibliography

Claesges, Ulrich. *Edmund Husserls Theorie der Raumkonstitution.* The Hague: Martinus Nijhoff, 1964.

Drummond, John J. "On Seeing a Material Thing in Space. The Role of Kinaesthesis in Visual Perception." *Philosophy and Phenomenological Research* 40 (1979–1980): 19–32.

Drummond, John J. *Historical Dictionary of Husserl's Philosophy.* Lanham: Scarecrow Press, 2008.

Gallagher, Shaun. "Body Schema and Intentionality." In *The Body and the Self*, edited by José Luis Bermudez, Naomi Elian, and Anthony Marcel, 225–244. Cambridge: MIT Press, 1998.

Gurwitsch, Aron. *The Field of Consciousness.* Pittsburgh: Duquensne University Press, 1964.

Gurwitsch, Aron. *Phenomenology and the Theory of Science*, edited by Lester Embree. Evanston: Northwestern University Press, 1974.

Gurwitsch, Aron. *Marginal Consciousness*, edited by Lester Embree. Athens: Ohio University Press, 1985.

Humboldt, Alexander von and Bonpland, Aimé. *Essay on the Geography of Plants*, edited by Stephen T. Jackson, translated by Sylvie Romanowski. Chicago: University of Chicago Press, 2009.

Husserl, Edmund. *Husserliana XI Analysen zur passiven Synthesis. Aus Vorlesungs- und Forschungsmanuskripten, 1918–1926*, edited by Margot Fleischer. The Hague, Netherlands: Martinus Nijhoff, 1966.

Husserl, Edmund. *The Crisis of European Sciences and Transcendental Phenomenology: An Introduction to Phenomenological Philosophy*, translated by David Carr. Evanston: Northwestern University Press, 1970.

Husserl, Edmund. *Husserliana XVI, Ding und Raum. Vorlesungen 1907*, edited by Ulrich Claesges. The Hague, Netherlands: Martinus Nijhoff, 1973.

Husserl, Edmund. *Thing and Space: Lectures of 1907*, translated by Richard Rojcewicz. Dordrecht: Springer, 1998.

Husserl, Edmund. *Analyses Concerning Passive and Active Synthesis: Lectures on Transcendental Logic*, translated by Anthony Steinbock. Dordrecht: Kluwer, 2001a.

Husserl, Edmund. *Husserliana XXXII, Natur und Geist: Vorlesungen Sommersemester 1927*, edited by Michael Weiler. Dordrecht: Kluwer, 2001b.

Husserl, Edmund. *Husserliana Materialienband IV, Natur und Geist: Vorlesungen Sommersemester 1919*, edited by Michael Weiler. Dordrecht: Kluwer, 2002.

Konopka, Adam C. "The Environed Body: The Lived Situation of Perceptual and Instinctual Embodiment." *Studia Phaenomenologica* 12 (2012): 289–308.

Konopka, Adam C. "Embodiment and *Umwelt*: A Phenomenological Approach." *Social Imaginaries* 1, no 2 (2015): 53–72.

Merleau-Ponty, Maurice. *Phenomenology of Perception*, 2nd ed., translated by Colin Smith. New York: Routledge, 1958.

Newton, Issac. *Philosophiae Naturalis Principia Mathematica*, 3rd ed., edited by Alexandre Koyre, Bernard Cohen, and Anne Witman. Cambridge: Harvard University Press, 1972.

Rynasiewicz, Robert. "By Their Properties, Causes and Effects: Newton's Scholium on Time, Space, Place and Motion. Part I: The Text." *Studies in History and Philosophy of Science* 26 (1995a): 133–153.

Rynasiewicz, Robert. "By Their Properties, Causes and Effects: Newton's Scholium on Time, Space, Place and Motion. Part II: The Context." *Studies in History and Philosophy of Science* 26 (1995b): 295–321.

Sartre, Jean Paul. *Being and Nothingness*, translated by Hazel E. Barnes. New York: Simon and Schuster, 1984.

Zahavi, Dan. "Husserl's Phenomenology of the Body." *Etudes phénoménologiques* 10 (1994): 63–84.

Conclusion

These investigations into the logics of habitat associations in early 20th century accounts of plant succession have been attempts to reconstruct some of the basic assumptions and explanatory principles at work in the science of plant ecology. Implicitly at work in the classifications of habitat associations of Henry Chandler Cowles' and Frederick Clements' respective accounts of plant succession were basic assumptions concerning plant forms and explanatory principles regarding the geographical distribution of plants. The identification of these philosophical differences in early 20th century plant ecology deepens and extends Ronald Tobey's historical claim that there were two main transatlantic traditions that significantly contributed to the dawn of American plant ecology – Alexander von Humboldt's physiognomic plant geography and Eugene Warming's physiographic ecology. Both Cowles and Clements (among others such as Henry Gleason and Arthur Tansley) were self-consciously aware of the different assumptions and principles of these contributions and vigorously debated them in light of their field research. These debates provide a unique point of entry into some of the theoretical issues in the philosophy of ecology and clarify some of the methodological principles that are operative in population and community ecology today.

The philosophical differences between Cowles' Chicago school and Clements' Nebraska school can be illustrated through the contrasting associational logic in their conception of plant collectives. As we have seen, these logics of habitat associations were developed in the classification systems designed to explain plant succession – the process in which one collective of interspecific plant populations characteristically changes into another. We have seen that Clements' account of plant succession involved a directional series of characteristic plant associations ("serial formations") that culminate in a climax community. A climax community, for Clements, was an association of plant populations, e.g., prairie grasses, that was optimally adapted to regional habitat variations, e.g., surface geology, altitude, precipitation, and temperature, through biological processes proper to the plant association as a collective whole. The plant association as a large-scale self-organization with biological processes proper to the whole achieves an equilibrium with its habitat. The collective is self-organizing in an analogous way as an individual organism in the Kantian sense is self-organizing. Clements succinctly states this basic thesis in the opening of *Plant Succession*, "As an organism, the formation arises, grows, matures, and dies."[1] In short, the associational logic of Clements' account of prairie succession involved an analogy between individual organisms and plant associations as a collective with biological processes proper to itself as a whole. What is a "formation" in Clements' account of plant succession? We have seen that, for Clements, plant formations are regional associations of interspecific plant populations that characteristically develop through a directional process of serial stages culminating in a climax community. The whole has functional processes proper to itself as a self-organized biological system that contribute to a stabilization of population changes.

These genealogical investigations have examined the basic assumptions and explanatory principles at work in Clements' account of succession and classification of plant formations. In particular, the conception of organic form in Clements' logic of plant collectives is illustrative of an epistemological idealism that operates with a notion of associative synthesis that is univocal to biological individuals and collectives. This logic of associative synthesis illustrates a common analytic feature of 19th century German biology. This tradition has been spoken of in many ways, e.g., organicism, romanticism, German idealism, and naturalism, and supposes a conception of organic form that was most notably articulated in the second division of Kant's *Critique of Judgment*. I have argued that Humboldt's physiognomic conception of "plant forms" illustrates this Kantian conception of the organism and, in particular, the logic of synthesis proper to reflective judgment. More specifically, Humboldt was primarily interested in the overall associative unity proper to the category of totality. This logical reasoning, and the notion of associative synthesis that accompanies it, finds a significant philosophical justification in Kant's identification of the *a priori* synthetic principle of the faculty of judgment in the Third Critique. I have reconstructed some of the basic assumptions and explanatory principles of this Post-Kantian tradition and have illustrated several logical senses in which Clements' account of prairie succession supposes on a conception of organic form in the Kantian sense.

These investigations have also been attempts to critique the complex logic of plant associations that Humboldt supposed when he pioneered the modern science of plant geography. As we have seen in the second investigation, Humboldt's penchant for multi-disciplinary methodology generated a notion of plant associations that had botanical, organic, and spatially distributive attributes. This complex notion of plant forms – collectives of geographically distributed plants – could be scientifically known through botanical classifications (particularly leaf morphology), physiognomic descriptions of organic attributes, and the mathematical measurements of habitat variations. Humboldt's new and bold question concerned the geographical distribution of characteristic plant associations. How are plant collectives geographically distributed? This was the basic question of 19th century plant geography and the early American plant ecologists inherited this basic question and addressed it through the dynamic processes involved in plant succession.

Humboldt's question concerning the geographical distribution of plant associations was the theoretically operative question for both the Nebraska and Chicago schools of early American plant ecology, but Cowles' physiographic

classification of plant associations supposed a different conception of plant form –
a nutritive conception of form that operated with a logic of habitat associations
that was more akin to a cellular logic of nutritive function in the physiological
sense, than an organic logic in the Kantian sense. The basic categories of Cowles'
physiographic classification of plant associations, e.g., hydrophytic, mesophytic,
and xerophytic, concerned soil moisture tolerance and the topological features
associated with surface water drainage. These water variations were the primary
characteristics through which plant associations were physiographically classified –
we find groups of plants in typical associations in a way that is primarily conditioned by soil moisture tolerance and the soil erosion and deposition involved in
water drainage. Cowles' account of succession also supposed a directionality,
but in a different logical sense in which Clements attributed an organic development to plant collectives. Plant succession was a directional process, for Cowles,
not in an organic sense, but a topological sense. Post-glacial erosion and drainage basins were the habitat features that produced the directionality – the end or
telos of which was the elevation baseline for water drainage. While a river valley
is perhaps the clearest example of topological base leveling, Cowles' account
of Lake Michigan dune succession also illustrates the directionality involved in
this physiographic approach. Lake Michigan is gradually receding and exposing
new shoreline that becomes populated by vegetation. As one walks inland from
the shoreline, a series of characteristic associations that culminate in mesophytic
edaphic forests, e.g., upland oak-savanna and beech-maple-hemlock, lowland
cedars, white birch, and cherry trees, illustrate a successive sequence. The directionality – never uniform – is not primarily the result of a self-organized biological process that is proper to the plant collective, but a directionality that is proper
to the post-glacial history of the lake. The habitat co-variation in this example is
the recession of the lake and the primary succession of the exposed shoreline and
foredunes. The directionality of this necessary fitness is geographical – the lake
is gradually receding and someday will empty entirely and be populated by terrestrial vegetation. In short, Cowles' physiographical account of dune succession
illustrates how the topological and geographic features of a habitat organize the
process of succession.

    As we have seen in the initial genealogical investigations, the philosophical
difference between Clements' and Cowles' accounts of succession can be historically illustrated through a contrast between two European traditions – Humboldt's
botanical physiognomy and those who were applying the basic insights of experimental plant physiology to the distributive question of plant geography, e.g.,
Andreas Schimper and Warming. I have argued that this second tradition represents a new decisive breakthrough in the history of early plant ecology. This
breakthrough introduced an alternate logic of habitat associations that supposed
a nutritive conception of plants. My claim is that advances in 19th century plant
physiology introduced a nutritive conception of plants that is logically different
from organic forms in the Kantian sense. These investigations have been primarily preoccupied with an immanent critique of the Kantian and Post-Kantian
assumptions and explanatory principles in plant ecology. Generally speaking,

I have problematized the botanical, organic, and spatial features of Humboldt's complex notion of form and have identified and clarified how a phenomenological theory of unified definite manifolds and part-whole logic of fitness are distinctive from standard approaches to multi-level explanations. I have not, however, sufficiently clarified the part-whole relations that are specifically operative in the nutritive conception of plant growth introduced by 19th century plant physiology. In conclusion, I would like to briefly sketch some of the basic assumptions and explanatory principles of this physiological tradition in early plant ecology. What were the essential insights of the 19th century plant physiologists that were applied to the question concerning the geographical distribution of plants? How do the modular structures of plant nutrition manifest part-whole relations that are different than those found in animal physiology? How did this nutritive conception of plant forms inform the new beginning at work in physiological approaches in plant ecology? In what sense did the field research of the methodologically rogue ecologists necessarily supplement the experimental design of the physiologists' laboratory?

The logic of habitat associations in physiographic plant ecology applies a nutritive conception of plant form in the physiological sense to the question concerning the geographical distribution of plants. This nutrient conception of plant form can be distinguished from the morphological conceptions of plant form as sensible shape in botanical taxonomies and romantic conceptions of plant forms as organisms in the Kantian sense (self-organized in wholes with heterogeneous parts associated contingent, purposive, and serially sequenced means-end relations). A nutritive conception of plant form in the physiological sense supposes a cellular logic of microscopic parts and wholes. This molecular logic of cellular associations involves both a morphological and physiological conception of cell forms. Morphologically speaking, plant cells are wholes with sensible shapes that can be distinguished into three concentrically nested parts – cell wall, protoplasm, and (in most cases) nucleus. These morphological parts of cellular wholes operate with a conception of form as sensible shape. Similar to botanical taxonomies that classify the sensible shapes of phenotypic characteristics, morphological accounts of cells and cellular associations examine the microscopic parts of cells in terms of sensible shapes. The obvious difference is that the sensible shapes of the parts of cells are microscopically amplified and the obvious similarity is that both botanical sensible shapes and cellular sensible shapes employ a notion of form that is proper to visual perception in the first instance. The methodological and technical breakthrough that makes this morphological conception of cellular form as sensible shape was Julius Sachs' insistence that the descriptions of microscopic observations are a legitimate source of primary data in the botanical sciences.

Cells are morphological wholes whose heterogeneous parts are not only visible to the microscopic eye as sensible shapes, but are also materially distinguishable in chemical composition. In addition to this static conception of morphological form (and presupposing it) is a physiological conception of nutrient form as the dynamic actualization of potential. Plant cells are morphological wholes with heterogeneous parts that have regular patterns of change throughout the phases of growth. The actualization of potential in the growth and development of plant cells can be spoken of in many ways according to the life-cycle of the plant tissue. The physiological sense of the growth and development of plant cells concerns the actualization of potential of nutrient production. A nutritive conception of plant form arises through the thematization of the growth of individual plants (rather than an intergenerational species as the ultimate end of the actualization of nutritive potential). Form in this physiological sense is the mature individual that successfully converts solar energy into starches through a process of photosynthesis. This actualization of nutritive potential is evident to the plant physiologist in at least two ways: 1) the chemical expression of morphology at various phases of cellular growth, and 2) the phenotypic characteristics of a mature individual. The physiologist makes functional claims in light of this conception of form. There is thus a compound notion of form in the cellular logic of the nutritive function of plant cells that includes both a morphological (static sensible shape) and physiological (dynamic actualization of potential) sense. In a contemporary context, this logic of nutritive form is expressed in terms of autotrophy.

This conception of nutritive forms is different from organic forms in the Kantian sense in a variety of ways. Perhaps an initial point of entry into this difference can be illustrated through a contrast between animal and plant physiology. Animals have organs (organic parts) that are heterogeneous, non-regenerative, and serially sequenced. Consider the organs involved in animal digestions, e.g., stomach, liver, kidney, and bladder. This digestive system is a serial sequence of heterogeneous and specialized organs that are non-regenerative. Not only are individual organs non-regenerative, e.g., new kidneys do not replace old ones, but the loss of one organ function leads to the loss of the whole digestive system. Each organ is functionally essential and non-iterative in a serial sequence. By contrast, plant physiologists operate with a different theory of unified definite manifolds and part-whole logic. My claim is that a nutritive conception of plant form in the physiological sense supposes a modular morphology that is regenerative, iterative, and comprised of non-serially sequenced or modular assemblages. In what follows, I briefly highlight some of the logical features of this conception of form and explore some of the differences between animal and plant physiology.

A nutritive conception of plant form in the physiological sense can be further identified and clarified through a reflection on the basic fact that plants are sessile – they do not locomote from one place to another. They have evolved in the Darwinian sense through a variety of adaptive strategies, such as the regeneration of morphological parts, e.g., leaves and bark, that are eaten by herbivores. Plants have a variety of adaptive strategies developed in trophic associations with herbivores, e.g., seed dispersal, rapid shoot growth, and leaf regeneration. Consider leaf regeneration in more detail. Plants overproduce the starches that are stored in leaves and can regenerate this nutritive content in the event that it is eaten by an herbivore. This regenerative storage process in autotrophic plants can be expressed in the cellular logic of nutritive forms. Consider how the process of plant cell division has a modular organization. By contrast to the serially sequenced and non-regenerative organs involved in animal physiology, e.g., a digestive system of heterogeneous and specialized organs, the cellular logics of the plant physiologist operate with a modular conception of regenerative, iteratively replaceable, and non-serially sequenced cellular assemblages. Plant function in this physiological sense is not related to specialized organs – plants do not use lungs to breath, mouths and stomachs to obtain nutrients, or skeletons to stand erect.[2] The cellular logic of the plant physiologist operates with a notion of cellular association that is not compositionally arranged in the same way as heterotrophic animals, which is to say, plants avoid concentrating nutritive functions in heterogeneous and non-iterative parts. Rather, the cellular logic of the plant physiologist conceived of plant functions as distributed manifolds of cellular tissues (cellular assemblages or networks) that add onto each other independently and without centralization in morphologically heterogeneous and non-regenerative organs. The metaphor of building "legos" might be helpful here in a very limited and artifactual sense. Consider the way that the perceptual interceptors of plants are modularly networked in decentralized and distributed nodes. Or consider how the vascular tissues in plant stems, shoots, and trunks circulate water and nutrients without a morphologically discrete organ such as a heart. It is in these senses that the cellular logic of nutritive forms can be distinguished from the logic of organic form proper to animal physiology.

This nutritive conception of plant form in the physiological sense was one of the basic assumptions of the late 19th century and early 20th century physiographers, occasionally referred to as "field physiologists." More specifically, physiographers such as Warming and Cowles classified habitat associations in a way that implicitly assumed a nutritive conception of plant form in the physiological sense. As we have seen in these investigations, the logic of habitat associations in these classifications of particular geographical places focused on the way vascular systems and growth rates of shoots contribute to water assimilation and transfer. The conception of nutrient form in the physiological sense assumed this thematic focus and contributed to the classification of plant associations in the various accounts of plant succession.

The specific geographical habitats of the field research in which Warming and Cowles pioneered physiographic principles included the sand dunes of the Northern European shore and the freshwater dunes of Lake Michigan. As we have seen, these sand dunes are among the most inhospitable habitats for plants due to the topological instability and extreme annual temperature variation. Almost nothing grows on a wandering dune. Dune grasses are geographically distributed on sand dunes through modular associations that collectively stabilize dune topography in assemblages of rhizome root structures and iteratively distributed leaf shoots with growth rates that can outpace the rates of topological variations of dune drifts. The pioneering grasses colonize the dune in ways that have a stronger metaphorical similarity with a colony of bees or ants, rather than an individual animal organism. These grass associations are networks with collective accomplishments that iteratively distribute physiographic processes, e.g., topological stabilization. Each blade of the dune grasses accomplishes topological stabilization in a similar way – by catching grains of predominantly quartz sand blown by the wind, vertically growing faster than the sand accumulation, and extending subsurface rhizomic root structures. These physiographic functions are iterative to the individual members of the assemblage, but are nevertheless jointly actualized and thereby accomplished collectively. The sense of the whole here is not as a second thing over and above the parts, but a collective with unifying moments that are immanent to them. These plant associations organize the topological variations of dune formations and, in turn, are modularly organized as iterative individuals by the habitat. In short, the logic of habitat associations in physiographic approaches to the question concerning the geographical distribution of plants operate with a nutritive conception of plant form in the physiological sense.

Allow me to briefly conclude with a few qualifications to this sketch of the basic assumptions and explanatory principles at work in the application of this nutrient conception of plant forms to physiographic logics of habitat associations. No individual habitat is the same as any other, which is to say, the notion of habitat is first and foremost ideographic. There are many kinds of particular habitats that are populated by various kinds of plant associations that do not all have the same logical features. Coral reefs, tropical forests, and inland lakes, for example, have different logics of habitat associations – each one must be empirically worked out in their own right. Not all plant associations have the same modular design and, moreover, physiographic accounts of plant succession are not the only tools in the plant ecologist's toolbox. In these physiographic senses, plant ecology is a geographical science in the first instance and thoroughly.

A physiographic ecology is also a physiological science in a qualified sense that can be clarified through a response to a final objection. The previous sketch of a cellular logic of plant forms and physiographic logic of habitat associations runs the risk of confusing several levels of description – the cellular, phenotypic, and geographic. The "confusion of levels problem" that haunts standard approaches to theories of multi-level explanations might at first glance also apply to the previous sketch. The objection could be succinctly posed: the growth of cellular plant tissue on a microscope slide or petri dish is different than the phenotypic grasses that populate a sand dune, and the abiotic features of the dune is another matter entirely. The logic does not slide from one level of description to another. The cellular logic of nutritive forms in the physiological sense is only metaphorically or analogically applied to plant associations on sand dunes, coral reefs, and forests. To superimpose the logic of cellular organizations to the phenotypic traits of plants is to succumb to the tendency of idealizing collective properties as hypostasized and secondary substances with physiological functions proper to itself. According to the line of reasoning in this objection, the logic of habitat associations in the physiographic sense commits the same error as a Kantian account. Rather than analogizing plant associations with organisms in the Kantian sense, physiography idealizes the cellular logic of the nutritive forms in the physiological sense in its application to plant associations. While the conception of plant form may be different in each case, the process of idealization is similar. According to this objection, physiography is also susceptible to the "tendency toward idealization" that is involved in the logic of habitat associations.

My response to this objection can be succinctly put: the confusion of levels problem in multi-level explanations is one of the reasons that a science like plant ecology needs a theory of unified definite manifolds and part-whole logic. Moreover, these investigations have illustrated how phenomenological methodology provides attractive theoretical resources to address this problem and deserves a seat at the table of the philosophy of science more generally. More to point, a phenomenological approach to this problem can be clarified through posing the fundamental question – what is a physiographic thing? My phenomenological claim is that a physiographic thing can be best formally understood as a unified definite manifold that is internally unified with (not externally related to) habitat associations that are organized in symmetrical part-whole relationships. More concretely, a physiographic thing is a phenotypic individual, e.g., a blade of dune grass, that changes in and through the changes in its habitat. The "real actuality" here is the plant given in the flesh in the everyday pre-theoretical contexts of gardening, hiking, and traveling. When physiographers theoretically reflect on the habitat relationships of phenotypic individuals, they do not represent levels of explanation that are somehow ontologically distinguishable from the real actuality of the physiographic things themselves. The physiographer does not need to be preoccupied with ontological concerns associated with multi-level causal explanations, but can appeal to a phenomenological shift in explanatory emphasis that neutralizes this preoccupation and shifts the explanatory burden to a descriptive account that tracks the necessary relationships among particular habitat associations. Physiographic things are phenotypic individuals considered in an irreducible habitat relation with an immanent logic that does not scale up in a hierarchy of generalizations. The relationship between phenotypic individual and habitat is indexed as the fundamental datum and the determinate logic of habitat associations tracks the particular co-variations of this relationship in search for necessary dependencies proper to the physiographic sense of the thing.

This phenomenological shift in theoretical interest not only frees the physiographer from the latent ontological presuppositions that burden standard approaches to theories of multi-level explanations, but opens a distinctive way of describing the collective achievements of plant associations. A physiographic approach to plant associations does not misapply a small-scale cellular logic of nutritive plant forms in the physiological sense to the large-scale organizations at work in the habitat fitness of plant associations. The logic does not slide among levels, but is specific to the empirical datum to both the cellular and phenotypic levels of description. The differences between the plant physiologist in the laboratory and the physiographer in the field can be simply clarified through the abstraction that the laboratory affords. The cellular logic of the plant physiologist reduces the phenotypic-individual-in-a-habitat to its basal conditions in an effort to isolate the variations that condition nutritive growth and thereby abstracts from this fundamental datum. Through this process of abstraction, the plant physiologist discovers essential insights into the cellular logic of nutritive plant growth. This second order justification crystalizes the distinctive features of plants as cellular assemblages that add onto each other independently and with distributed nutritive functions that are not centralized in morphologically heterogeneous and non-regenerative organs. While the logic of cellular assemblages and phenotypic plant associations remains tied to the specific domain of empirical phenomenon, the essential insight into plants as distinctive nutritive forms is operative at both levels. Physiographers appropriate this conception of nutritive plant forms in the physiological sense as the leading clue in their investigation of the logic of habitat associations that are specific to geographical places. In short, a nutritive conception of plant form is the basic assumption for both the physiologist and physiographer, but the physiographer employs this assumption in the phenotype-in-a-habitat that originates in a pre-theoretical attitude as a fundamental datum. While the conception of nutritive form slides among the levels of description, the logic remains immanently tied to the respective levels. With this qualification in mind, a physiographic approach to the question concerning the geography of plants operates with a nutritive conception of plant form in the physiological sense in an attempt to identify and clarify the characteristic ways in which habitats condition where plants grow.

Notes

1 Frederick Clements, *Plant Succession: An Analysis of the Development of Vegetation* (Washington: Carnegie Institution, 1916).
2 See Stefano Mancuso and Alessandra Viola, *Brilliant Green: The Surprising History and Science of Plant Intelligence*, 2nd ed., trans. Joan Benham (Washington D.C.: Island Press, 2015), 34.

Index